Complexities of Production and Interacting Human Behaviour

Yuji Aruka
Editor

Complexities of Production and Interacting Human Behaviour

Physica-Verlag

Editor
Professor Dr. Yuji Aruka
Chuo University
Faculty of Commerce
Higashinakano 742-1
Hachioji, Tokyo 192-0393, Japan
aruka@tamacc.chuo-u.ac.jp

ISBN 978-3-7908-2617-3 e-ISBN 978-3-7908-2618-0
DOI 10.1007/978-3-7908-2618-0
Springer Heidelberg Dordrecht London New York

Library of Congress Control Number: 2011924880

Cover design: WMXDesign GmbH, Heidelberg, Germany

Printed on acid-free paper

Physica-Verlag is a brand of Springer-Verlag Berlin Heidelberg
Springer-Verlag is a part of Springer Science+Business Media (www.springer.com)

Preface

This book is a collection of my essays on economic theory and its related issues. The collected papers explore complex production systems and heterogeneously interacting human behaviour, my keen pursuits these days. In celebration of my 60th birthday I have decided to compile my previous works into a single book. Most of the chapters are published papers. The first chapter was added in order to suggest a new perspective for analyzing the socio-economic system and interacting human behaviour. In my previous edited book *Evolutionary Controversies*, I described a similar perspective on behalf of the Japan Association for Evolutionary Economics (JAFEE): "Toward a new transdisciplinary approach for evolutionary controversies". This publication was internationally successful. After its publication I attended and hosted several international conferences and workshops, and have learned new perspectives. Thus the first chapter reflects my current view, inspired by my recent activities.

This book is in three parts: Part I, Complexities of production and social interaction; Part II, Moral science of heterogeneous economic interaction; and Part III, Avatamsaka's dilemma of the two-person game with only positive spillover.

Part I covers a lot of ground from production to social interaction. In this preface, I will just refer to the topic of entropy in production. Professor Jürgen Mimkes (University of Paderborn), my collaborator, suggested the significance of entropy in a productive–economic cycle. In Chaps.1 and 4, I try to make the application of the idea of entropy to my economic system brighter in light of Mimkes's assistance. In Part III, I discuss the topic of the Avatamsaka game, in which any defector may use his rival's cooperation for his gain. Without the rival's cooperation, however, his gain cannot be guaranteed. The expected value of a gain for any agent could be reinforced if it were generated by a higher average rate of cooperation, as measured by the frequency of mutual cooperation. Here there is a macroscopically weak control mechanism, but not the personal mutual fate control described by Thibaut and Kelley (in Chap.12). My continuing interest in this game was supported by Professor Eizo Akiyama (Tsukuba University), my collaborator, who discovered a new result in 2006. I heartily thank Professors Mimkes and Akiyama for their insightful assistance and hospitality.

As to the context of the book, Part II in particular, it may be helpful for readers to know my short research itinerary. I began my career studying the subjects of Part I. Strictly speaking, this was inspired by the study of Piero Sraffa in the course of doctoral studies under the late Professor Izumi Hishiyama at Kyoto University. He taught me how to grasp a certain basis of Piero Sraffa's work as well as the moral philosophy of John Maynard Keynes, a sponsor of Sraffa. With his supervision, I became capable of simultaneously studying both production systems and moral science. The analysis of production and moral science, the main subjects of the first two parts of this book, was thus inspired.

I would also like to gratefully refer to some special professors that influenced Part I. I am very much indebted to Professor Bertram Schefold (Goethe University Frankfurt). Without learning his brilliant contributions on the Sraffa–von Neumann joint production system my research and idea could never have emerged at all. I must also be grateful to him for suggesting many intelligent insights, in particular, the German idealism of Stefan George and in German historical school of economics. For a mathematically strict view on Sraffa I was much influenced by Professor Yoshinori Shiozawa (Osaka City University and Chuo University). Taking a broader view of economics, I learned a great deal about an integrative approach to linear production systems and nonlinear/chaotic dynamics from the late Professor Richard Goodwin (University of Siena and Peter House, Cambridge). His insight was quite helpful for me to shift to a study of complex systems during the 1990s.

The last decade of the last century was really an exciting period in science. The so-called revolution of nonlinear science as the key node (hub) provided us with a general link to various fields of science that eventually led to an interdisciplinary or transdisciplinary method as a standard of science. Fortunately, it was easy enough for the social sciences to find many chances to wire the key node. This advent of a new method of science often gave us new opportunities for discussions with scholars of different fields. At the beginning of this century, we arranged JAFEE2000, the international conference held in Tokyo in 2000, along with *Evolutionary Controversies*, its publication of selected papers. This conference helped internationalize JAFEE's activity as well as my own. Soon afterwards, in 2004, JAFEE decided to issue its own international journal, *Evolutionary and Institutional Economics Review*. I was appointed Editor-in-Chief in 2006. My commitment to this editorial work has currently become a new link for my integrative studies.

Another fortunate turn for my own studies was my commitment to the group Workshop for Economics with Heterogeneously Interacting Agents (WEHIA). We can hardly discuss WEHIA without referring to Professor Mauro Gallegati (Universita' Politecnica delle Marche), who was sincerely devoted to organizing this group. This group has now evolved as the Society for Economic Science with Heterogeneous Interacting Agents (ESHIA), a more formal society. Professor Akira Namatame (National Defense Academy of Japan) was decisively influential in arranging the new organization by undertaking the secretary's office. Besides the activities of WEHIA/ESHIA, Professor Namatame has currently been giving me a broader perspective and useful connection as a very good partner, particularly in the field of information/computer sciences. I must cordially thank Professors Gallegati and Namatame for their continued cooperation.

In the first decade of this century I have become intrigued by a promising grand view for reconstructing economics, based on the idea of a master equation used in statistical physics. A seminal approach for this idea is synergetics, coming from laser optics. Its application to the social sciences was ingeniously arranged as sociodynamics by Professor Wolfgang Weidlich (University of Stuttgart). The author kindly appointed me as a translator of his great book *Sociodynamics*. Thus I had the opportunity to learn his grand vision closely. Another great opportunity was given to me by Professor Masanao Aoki (UCLA). I happened to obtain a joint research position for him in my university. We held a socio-econophysics conference, the first one in Japan, in 2003. Professor Aoki taught me not only about the importance of stochastic process but also a new innovative framework to embrace unknown agents as well as non-self averaging properties. His theory is a remarkably attractive set of devices for future studies on economics. My recent direct communications with these two great scientists was one of my greatest pleasures. The final chapter of this book, in Part III, resulted from discussions with Professor Aoki, while several chapters of Part II were based on the sociodynamic approach in Professor Weidlich's work.

The idea of socio-econophysics has been explicitly arranged by the socio-econophysics circle inside the German Physical Society (DPG) at the end of last century. In this circle Professor Dirk Helbing and Professor Frank Schweitzer were critical. I had the opportunity to invite them to Japan in 2002 and 2003, respectively. They have recently developed CCSS, organized in ETH Zurich. Thanks to Professor Helbing, I had time to stay at ETH in February 2010 and had many useful discussions.

It is always exciting for me to find a scholar who is interested in subjects other than his own. Nothing gives me greater pleasure than to meet and discuss many things with Professor Bertrand Roehner (University of Paris), not only one of the pioneers of econophysics but also a careful observer of the effects of institutional factors. His method may be called institutional econophysics. His interests also cover historical ground, including Japan, my country. His contributions will be quite beneficial to me in future study.

In my retrospection at the beginning of my study I originally tried to strive for a new method for economics. The first seed for escaping from mainstream thought seems to have been planted since when I learned the German economics of Erich Schneider (University of Kiel and Institute for World Economy, Kiel), and of Heinrich von Stackelberg (University of Bonn) by the late Professor Yoshio Yamakawa and Professor Tatsuji Owase (Waseda University), who were my professors in graduate school. For instance, von Stackelberg's idea of leadership in duopolistic markets continues to be popular even in economic journals. The introduction of leadership implies the introduction of feedback effects among the interacting agents. The effects may thus offer a new strategic stage separate from the original arena for competition. Consequently, agents may be involved in restructuring a system. As Pu-yan Nie proved, the introduction of alternating leadership could be interpreted in terms of bi-level optimization. The new domain thus spanned is much broader than that of a standard Cournot–Nash framework, in which agents are forced to miss their underlying basis. On the other hand, agents involved in restructuring a system of necessity must be socio-economical because an economy always stands on society.

Many economists stubbornly keep standing on a restrictive domain arranged by the mainstream, though the real world is rapidly becoming more and more complicated. We cannot help but guess at a serious poverty of philosophical thought in the attitude of many such economists, who should be recommended to a complexity of thinking as in the works of Professor Klaus Mainzer (University of Technology Munich). Due to his works, as well as through direct dialogue with him, I have realized a so-called integration of British empiricism and German idealism. Faced with the great mess of the world, the idea of absolute goodness deserves to be urgently reconsidered. This is the reason why I added a supplemental chapter (as the Appendix) on a book review of Mainzer (2008) at the end of this book.

My view is merely a small consequence of the many intellectual discussions with my colleagues. I have already mentioned some of them, but I would still like to express my gratitude to other important colleagues. I will limit my expressions only to those who have affected me directly. They were and are the source of my motivation and inspiration. Dispensing with titles: Izumi Hishiyama for Sraffa and Cambridge economics; Yoshio Yamakawa and Tatsuji Owase for German economics; Bertrand Schefold for Sraffian mathematical systems; Richard Goodwin for nonlinear and chaotic economic dynamics; Wolfgang Weidlich and Masanao Aoki for the master equation and its applications; Bertrand Roehner for institutional econophysics; Klaus Mainzer, Peter Erdi, and Ping Chen for complexity thinking; Dirk Helbing, Frank Schweitzer, Mitsugu Matsushita for socio-econophysics; Staoshi Sechiyama, Yoshinori Shiozawa, Kiichiro Yagi, Hiroshi Deguchi for JAFEE; Mauro Gallegati, Akira Namatame, Thomas Lux, Enrico Scalas, Shu-Heng Chen for ESHIA; Ulrich Witt, Geoffrey Hodgson, Carsten Herrmann-Pillath for evolutionary economics; Stefan Guastello and Barkley Rosser, Jr. for the Society for Chaos, Psychology, and Life Sciences; Sobei Hidenori Oda for experimental economics; Jun Tanimoto for evolutionary games; and Jürgen Mimkes and Eizo Akiyama as my collaborators.

At Springer, I am indebted to Dr. Christian Caron, Executive Editor, and Ms. Barbara Fess, Senior Editor, Springer Heidelberg. Without their kind hospitality this book would never appear at all. Their decision to publish it is a genuine honor for me.

Finally, the original sources for the papers in this collection should be listed. Except for the opening chapter dedicated for this book "A perspective for analyzing the socio-economic system and interacting human behaviour," the remaining papers were published previously. I would like to thank the publishers who so kindly gave permission to reproduce them in this book.

Part I. Complexities of Production and Social Interaction

Chapter 2. Aruka, Y. (1991) 'Generalized Goodwin theorems on general coordinates, structural change and economic dynamics', 2(1), 69–91; repr. in

J. C. Wood (ed.) (1996), *Piero Sraffa: Critical Assessments, IV* (London: Routledge), 127–153.

Chapter 3. Aruka, Y. (2000) 'Possibility theorems on reswitching of techniques and the related issues of price variations', *Bulletin of the Institute of Economic Research, Chuo University*, 30, 79–119.

Chapter 4. Aruka, Y. and Mimkes, J. (2006) 'An evolutionary theory of economic interaction: introduction to socio- and econophysics', *Evolutionary and Institutional Economics Review*, 2(2), 145–160.

Chapter 5. Aruka, Y. (2001) 'Family expenditure data in Japan and the law of demand: a macroscopic microeconomic view' in H. Takayasu (ed.) (2001), *Empirical Science of Financial Fluctuations: Proceedings of the Nikkei Econophysics Symposium* (Tokyo: Springer), 294–303.

Chapter 6. Aruka, Y. (2004) 'How to measure social interactions via group selection? A comment: cultural group selection, co-evolutionary processes, large-scale cooperation', *Journal of Economic Behavior and Organization*, 53(1), 41–47.

Part II. Moral Science of Heterogeneous Economic Interaction

Chapter 7. Aruka, Y. (2004) 'Exploring the limitations of utilitarian epistemology to economic science in view of interacting heterogeneity', *Annals of the Japan Association for Philosophy of Science*, 13(1), 27–44.

Chapter 8. Aruka, Y. (2007) 'The moral science of heterogeneous economic interaction in the face of complexity', in Theodor Leiber (ed.) *Dynamisches Denken und Handeln Philosophie und Wissenschaft in einer komplexen Welt, Festschrift für Klaus Mainzer zum 60* (Stuttgart Geburtstag S. Hirzel Verlag), 171–183.

Chapter 9. Aruka, Y. (2008) 'The Evolution of moral science: economic rationality in the complex social system', *Evolutionary and Institutional Economics Review*, 4(2) vol. 4, no. 2, 217–237.

Part III. Avatamsaka's dilemma of the two-person game with only positive spillover

Chapter 10. Aruka, Y. (2001) 'Avatamsaka game structure and experiment on the web', in Y. Aruka (ed.), *Evolutionary Controversies in Economics* (Tokyo: Springer), 115–132.

Chapter 11. Aruka, Y. (2001) 'Avatamsaka game experiment as a nonlinear Polya urn process', in T. Terano, A. Namatame et al., *New Frontiers on Artificial Intelligence* (Berlin: Springer), 153–161.

Chapter 12. Aruka, Y., and Akiyama, E. (2009) 'Non-self-averaging of a two-person game with only positive spillover: a new formulation of Avatamsaka's dilemma', *Journal of Economic Interaction and Coordination*, 4(2), 135–161.

Appendix

Y. Aruka, review of Klaus Mainzer, *Der kreative Zufall: Wie das Neue in die Welt kommt (The Creative Chance: How Novelty Comes into the World, (In German))* (München: C.H. Beck, 2007) in *Evolutionary and Institutional Economics Review*, 5(2) (2008), 307–316.

I received research support from the Japanese Society for the Promotion of Science (JSPS) and from my university. Several chapters were the products of these funds. I am indebted to the governing bodies for the Grant-in-Aid for Scientific Research of the Japanese Society for the Promotion of Science (JSPS) No.10430004, No. 14580486, and No. 18510134, and for Chuo University Grant-in-Aid 2006-7 and 2008, respectively.

At my studio in Otowa, Tokyo, on 5th May 2010

Yuji Aruka

Contents

Contributors

Eizo Akiyama Graduate School of Systems and Information Engineering, University of Tsukuba, Tennodai 1-1-1, Tsukuba, Ibaraki 305-8573, Japan, eizo@sk.tsukuba.ac.jp

Yuji Aruka Faculty of Commerce, Chuo University, Higashinakano, Hachioji, Tokyo 192-0393, Japan, aruka@tamacc.chuo-u.ac.jp

Jürgen Mimkes Department of Physics, The University of Paderborn, Warburger Str. 100, Paderborn D-33098, Germany, mimkes@zitmail.uni-paderborn.de

Chapter 1
A Perspective for Analyzing the Socio-Economic System and Interactive Human Behaviour

Yuji Aruka

We should distinguish freedom from randomness, particularly in a socio-economic system as Mainzer (2007) has systematically envisaged. As econophysics has successfully proved the properties of power law as well as lognormal distribution in an actual socio-economic system,[1] we also are faced with a much higher probability of unfair or nonegalitarian consequences for equal opportunities.[2] This belief of a free market system should be studied. We briefly look at a recent idea of complex networks to characterize a distribution of systems.

1.1 Mechanism for the "Rich Get Richer" Phenomenon

1.1.1 Complex Network Reciprocities

The recent development of network analysis since Barabási and Albert (1999) can successfully focus on how a network could generate a scale-free property. As the complex network analysis showed, a huge network system must be too vulnerable to collapse due to a certain strategic aggression. This kind of vulnerability has been seen as closely connected to the scale-free property. It is well known that econophysics successfully examined the scale-free property of the modern economic system in various aspects of society. The management of this property will be our subject. Therefore we examine a preferential attachment to a random network to examine whether the network evolves as a scale-free system or not. Here we

---

[1]The lognormal distribution is a reference point between a Gaussian distribution and a power law distribution. If the standard deviation were taken towards infinity, the lognormal distribution could exhibit a power law, while it could be a normal one if the standard deviation were too small. See Kuninaka and Matsushita (2009, p. 1148).

[2]See Caon et al. (2007). Also see Drăgulescu and Yakovenkoa (2000, 2001).

Y. Aruka
Faculty of Commerce, Chuo University, Higashinakano, Hachioji, Tokyo 192-0393, Japan
e-mail: aruka@tamacc.chuo-u.ac.jp

Y. Aruka (ed.), *Complexities of Production and Interacting Human Behaviour*,
DOI 10.1007/978-3-7908-2618-0_1,

employ the idea of assortative/dissortative mixing to foresee the result. Reciprocity is one of the important features of networking, and it is one that may be dominated by a type choice of preferential attachment. This analysis suggests management of complex networks. More importantly, our network modeling has an essentially equivalent system to a Polya urn process, which may also generate the increasing return or winner-take-almost process of a modern production system.

1.1.2 Monetary Economy

In a random monetary exchange, there is a certain interesting result from a very simple setting given by Wright (2005):

Monetary Exchange[3]

Step 0. Initially, there is an egalitarian distribution of individual monetary stock. Step 1. Choose two randomly selected agents and then transfer a certain amount of one's monetary holding to the other agent. Step 2. Refresh the last round and then repeat until a certain round.

Traders can start exchanging from their equal initial balances. In this simulation, there is initially an egalitarian distribution of monetary holdings: each of 500 agents all has 100. After merely 10 rounds of exchange, however, the system may shape a power law distribution of money holdings, then approaching a maximum value of entropy. That is to say, the monetary economy to be randomly exchanged will give rise to a power law distribution (Fig. 1.1).

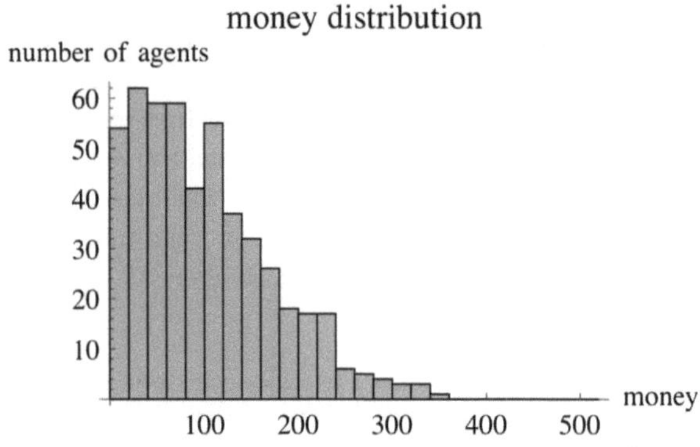

Fig. 1.1 Statistical mechanics of monetary exchange (1)

[3]"Statistical Mechanics of Money" from The Wolfram Demonstrations in Project http://demonstrations.wolfram.com/StatisticalMechanicsOfMoney/, contributed by Ian Wright, after work by A. A. Dragulescu, V. M. Yakovenko, and Justin Chen.

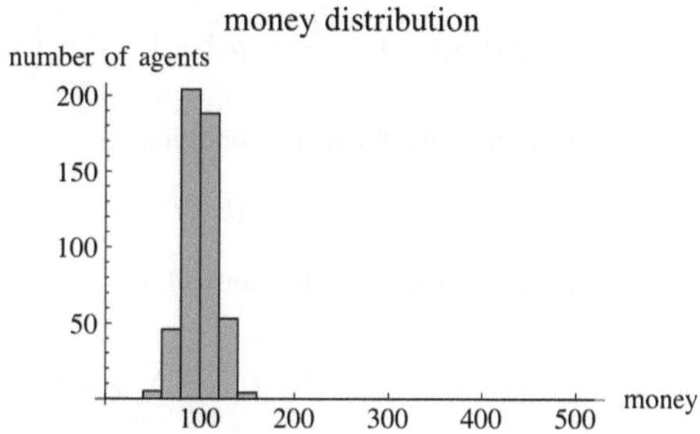

Fig. 1.2 Statistical mechanics of monetary exchange (2)

In this setting the random amount of money to be transferred may be any size. Next, we introduce a policy to regulate a size of the monetary transfer using a logarithmic transformation of the random amount of money. However, it is immediately apparent that only a small amendment on the control of agent numbers will give a drastically changed picture. Thus, without bigger transactions, we keep a relatively egalitarian state in this system even after 50 rounds of exchange. It is important that we never restrict "random exchange" (Fig. 1.2).

It is also interesting to analyze whether the number of agents may influence the result.

1.1.3 A Preferential Attachment with a Growing System

Barabási and Albert (1999) provided a rule to generate a power law distribution, though they were not successful in giving their desired property a "small world". Their rule is as follows:

BA Rule

Step 0. There is a perfect graph with m nodes K_m in an initially random network. Step 1. Add a new node to span a new link to one of the existing nodes with a probability $\pi(k_i)=k_i/\Sigma_i k_i$. Step 2. Iterate Step 1 until at a certain specified number is reached.

In this model, the average mean is

$$<k> = 2mt. \quad (1.1)$$

The probability that the node added to at the s-th step becomes degree k is $p(k,s,t)$; the probability that the node of degree k increases its own degree is $\frac{k}{2t}$. We then have the following master equation:

$$p(k,s,t+1) = \frac{k-1}{2t}p(k-1,s,t) + \left(1 - \frac{k}{2t}\right)p(k,s,t). \tag{1.2}$$

Here we impose the boundary condition[4]:

$$p(k,t,t) = \delta_{km}. \tag{1.3}$$

The degree distribution of the network is

$$P(k,t) = \frac{1}{t}\sum_{s=1}^{t} p(k,s,t). \tag{1.4}$$

Taking a sum of both sides from 1 to t on s, it holds:

$$(t+1)P(k,t+1) - tP(k,1) = \frac{1}{2}\{(k-1)P(k-1,t) - kP(k,t)\} + \delta_{km}. \tag{1.5}$$

Evaluated at a stationary state of $P(k) = P(k, t+1) = P(k, t)$, it follows:

$$P(k) + \frac{1}{2}\{kP(k) - (k-1)P(k-1)\} = \delta_{km}. \tag{1.6}$$

Hence it gives the solution:[5]

$$P(k) = \frac{2m(m+1)}{k(k+1)(k+2)}. \tag{1.7}$$

If we take a large number as k, it leads to bringing the power law[6]:

$$P(k) \propto k^{-3}. \tag{1.8}$$

Thus it is likely that the higher the degree of a node, the more the node connects with a node of higher degree. Hence we can find a reinforcement mechanism of the "rich get richer" phenomenon.

1.1.4 Preferential Attachment with a Nongrowing System

Instead of a case with growth, we can assume a network that never grows but reconnects mutually. Ohkubo and Yasuda (2005) and Ohkubo et al. (2006) did not

[4] $\delta_{ij} = 1$ for $i = j$; 0 for $i \neq j$.

[5] Since $P(m) = \frac{2}{m+2}$ and $P(k) = \frac{k-1}{k+2}P(k-1)$ for $k > m$, it follows the solution. This solution is verified if we obtain $P(m) = 2/(m+2)$ by setting $k = m$ in the equation for $P(k)$.

[6] See Krapivsky et al. (2000) and Krapivsky and Redner (2001).

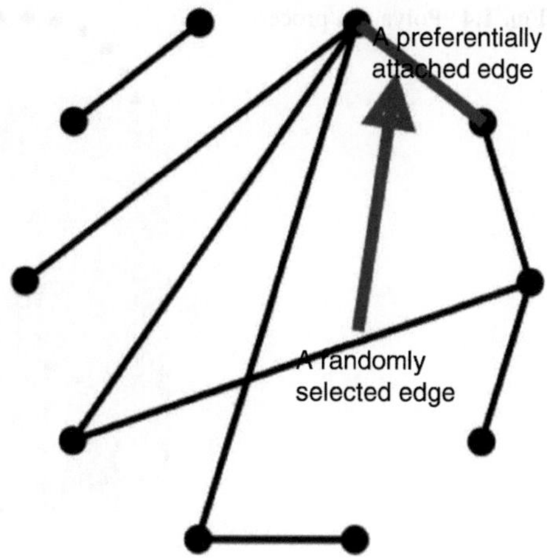

Fig. 1.3 Ohkubo rule

only propose a new rule for this nongrowing network but also proved that the rule was equivalent to a Polya urn process. We presume a fitness parameter distribution $\varphi(\beta)=\{\beta_i\}$[7] that is capable of creating a different possibility for network formation. The rule will be as follows:

Ohkubo rule

Step 0. There is initially given a random network of N nodes and M edges. Step 1. Specify randomly an edge l_{ij}. Step 2. Replace the edge l_{ij} by an edge l_{im} whose node m is chosen randomly with a probability $\Pi_m \propto (k_m + 1)^{\beta_m}$.

Here k_m is the degree of node m, and β_m a fitness parameter of node m.

According to Ohkubo et al. (2006), using a Polya urn process, we have a system where there are N urns and M balls (Fig. 1.3). We denote the number of balls in urn i by n_i. The total number of balls is denoted as

$$M = \sum_{i=1}^{N} n_i. \tag{1.9}$$

We then define the energy of each urn as

$$E(n_i) = -\ln(n_i!). \tag{1.10}$$

Hence the Hamiltonian of the whole system is

[7]The fitness parameters are assumed to be time independent.

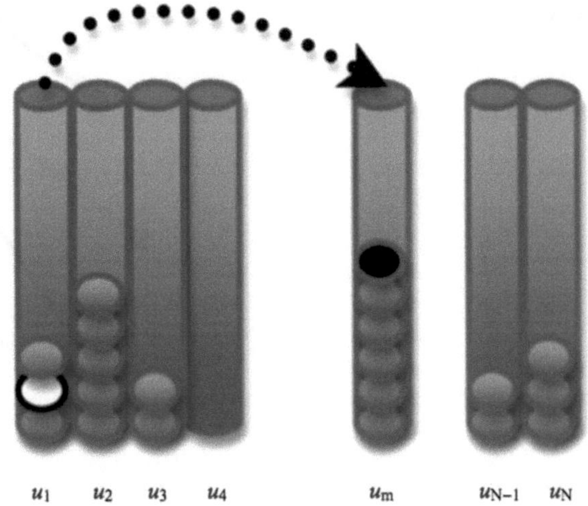

Fig. 1.4 Polya urn process

$$H = \sum_{i=1}^{N} E(n_i). \tag{1.11}$$

This definition of energy may lead to attaching the un-normalized Boltzman weight to urn i:

$$p_{n_i} = \exp[-\beta_i E(n_i)] = (n_i!)^{\beta_i}. \tag{1.12}$$

The use of the heat-bath rule gives the transition rate $W_{n_i \to n_{i+1}}$ from the state n_i to n_{i+1}.[8]

$$W_{n_i \to n_{i+1}} \propto \frac{p_{n_{i+1}}}{p_{n_i}} = \frac{(n_i + 1!)^{\beta_i}}{(n_i!)^{\beta_i}} = (n_i + 1)^{\beta_i}. \tag{1.13}$$

A Polya urn rule (Fig. 1.4)

Step 0. Initially, there is a partition of N urns (boxes) and M balls.
Step 1. Randomly draw ball n_{ij} in any box i.
Step 2. Replace it in an urn selected with a transition rate $W_{n_i \to n_{i+1}} = (n_{n_i} + 1)^{\beta_i}$.

In summary, we have the next equivalence (Table 1.1):

[8] See Drouffe et al. (1998).

Table 1.1 Equivalence between a network connection and a Polya urn process

	Network reconnection	Polya urn process
Step 0	N nodes and M edges	N urns and M balls
Step 1	Choose randomly an edge l_{ij}	Draw randomly a ball n_{ij} in any box i
Step 2	A new edge l_{im} with a probability $\Pi_m \propto (k_m + 1)^{\beta_m}$	A new urn with a transition rate $W_{n_i \to n_{i+1}} = (n_{n_i} + 1)^{\beta_i}$

It is also noted that we still keep randomness in networking until Step 2, though we imposed a stochastic preferential attachment rule at Step 3.

1.1.5 Assortative/Dissortative Mixing

In the argument above, we have seen a kind of assortative mixing. It is well known that in social networks, nodes having many connections are inclined to be connected with other highly connected nodes[9] while in technological and biological networks nodes of higher degrees are preferably attached with nodes of lower degrees.[10] The former property is called assortativity or assortative mixing. The latter is called dissortativity or dissortative mixing. If we apply the terms of degree correlation to our degree mixixng, we can state:

Mixing

Assortative mixing: Nodes of similar degrees mutually attract ⇒ A positive degree correlation.

Dissortative mixing: Nodes of higher degree tend toward nodes of lower degree ⇒ A negative correlation.

A different assortativity may generate a different network with different features. Here we like to specify "assortativity" by using a parameter p due to Xulvi-Brunet and Sokolov (2005).

Let a degree distribution which is linked to nodes of degree k be $P(k'|k)$. We denote the average nearest neighbor degree of nodes which is linked to a node of degree k by

$$\sum_{k'} k' P(k'|k). \tag{1.14}$$

Let E_{ij} be the probability that a randomly selected edge of the network connects any two nodes, one of degree i and the other of degree j; let the total number of links of the network be L. It then holds[11]:

[9] See Newman (2002) and Capocci et al. (2003).

[10] See Newman (2003) and Pastor-Satorras et al. (2001).

[11] In this article, we are dealing with undirected graphs. We assume $i \leq j$, without losing generality, since $E_{ij} = E_{ji}$.

$$\varepsilon_{ij} = \frac{E_{ij}}{L}. \tag{1.15}$$

When the probability of a link attached to a node of a certain degree is independent from the degree of the attached node, the network is then called uncorrelated if there is not any correlation between a pair of degrees of any two nodes (i, j):

$$\varepsilon_{ij}^r = (2 - \delta_{ij}) \frac{iP(i)}{<i>} \frac{jP(j)}{<j>}. \tag{1.16}$$

Operator $<i>$ means the first moment or expected mean (average) of i. Assortativity then means that nodes of similar degrees tend to be connected with a larger probability than in the uncorrelated case:

$$\varepsilon_{ii} > \varepsilon_{ii}^r \tag{1.17}$$

We thus define "assortativity"[12]:

$$A = \frac{\sum_i \varepsilon_{ii} - \sum_i \varepsilon_{ii}^r}{1 - \sum_i \varepsilon_{ii}^r}. \tag{1.18}$$

We define a variable F by using:

$$F_{ln} = \sum_{r=l} \sum_{s=r} E_{rs} \text{ for } r \leq s, l \leq n. \tag{1.19}$$

On each reconnection, F_{ln} either is increased by 1 or decreased by 1 or is unchanged. We denote f_{ln} by

$$f_{ln} = \frac{F_{ln}}{L}. \tag{1.20}$$

In detail, the respective rates of transition will be as follows:

Transition rates

$(X_{ln} - f_{ln}) + p(X_{ln} - f_{ln} + f_{ln,l-1})^2$	for $F_{ln} \to F_{ln} + 1$
$f_{ln}[(1-p)(1-2X_{ln}) + p(X_{1,l-1} - f_{1,l-1} - f_{ln}) + f_{ln}]$	for $F_{ln} \to F_{ln} - 1$

For simplicity, here we use the definition[13]

[12] See Xulvi-Brunet and Sokolov (2005).

[13] X is defined as the average of a degree distribution of connection $P_{\text{link}}(k) = kP(k)/k$.

$$X = \frac{\sum_{k=1}^{n} kP(k)}{<k>}. \tag{1.21}$$

It then obtains the following solution f_{ln}[14]

$$f_{ln} = \frac{X_{ln}^2 + (B_n - B_{n-1})^2}{\frac{1-p}{2} + pX_{ln} + B_n + B_{n-1}} \text{ for } l \leq n. \tag{1.22}$$

Here $B_n = \sqrt{\left(pX_{ln} + \frac{(1-p)}{4}\right)^2 - pX_{ln}^2 + \left(1 + \frac{p}{2}\right)}$

Hence it follows:

$$\varepsilon_{ij} = f_{ij} - f_{ij-1} - f_{i+1j} + f_{i+1j-1}. \tag{1.23}$$

Therefore, as seen from the definition of assortativity, A is a function of f and p:

$$A = A(f, p). \tag{1.24}$$

If $p = 0$, the network is fully uncorrelated, and then $A = 0$. If $p = 1$, the network is fully assortative, and $A = 1$ and A has a maximum ($A < 1$). Xulvi-Brunet and Sokolov (2005) verify that the average degree of the network never decreases as A increases by changing p.[15]

1.2 The System of Production and Heterogeneous Interactions

In the section above, we argued the evolution of network systems. There are broader applications of this evolution to various fields. One typical example of an application to the economic system is power laws. A certain economic concentration has a peculiar mechanism to generate its own network, not exceptional in a production system.

1.2.1 The Recycling Effect of Production

The idea of rewiring nodes in a network could be applied to production recycling. Holland (1992, Chap. 4; Holland 1995, pp. 23–26) pointed out that while it is not

[14]Xulvi-Brunet and Sokolov (2005, pp. 1434–1435).

[15]Figure 1 of Xulvi-Brunet and Sokolov (2005) validated this fact in the cases of $A = 0$ (uncorrelated network), $A = 0.26$, $A = 0.43$, $A = 0.62$ (maximal assortativity).

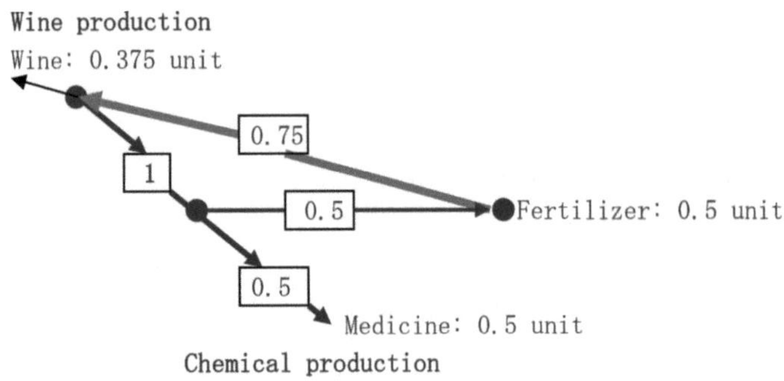

Fig. 1.5 A recycling of production

particularly surprising that recycling can increase output, its overall effect in a network with many cycles is striking. A tropical rain forest illustrates the point. In this context, Holland also cited the von Neumann economic growth model of production as a network. It is true that we can find a recycling mechanism in a von Neumann model von Neumann (1937). For instance, suppose that there is a three-commodity production system using wine, medicine, and fertilizer. Here each node • means basic product. The terminals of the arrows mean final consumption, as shown in Fig. 1.5.

This figure actually depicts a recycling system of production. In a real system of human production, each link of the recycling, i.e. each production process, must be currently associated with each labor input. If we remove an edge-node • within the recycle in Fig. 1.5, the system may degenerate into another one. This operation is called "truncation" and is just the same as the procedure of rewiring nodes. However, we will need a more sophisticated rule than the assortative/dissortative rule as argued above.

1.2.2 The von Neumann Economic Growth Model

Production with recycling is a consequence of interactions of heterogeneously different processes. We use a slightly revised version of the original von Neumann economic model from Schefold (1976, 1997). The detailed features of this model will be discussed in Chap. 2 of this book.

The economic system of production is comprised of dual systems of price and quantity. Production is achieved through a recycling network. In classic economics, we call this type of production "inter-industrial reproduction". For a brief illustration, production of the outputs by the recycling of themselves is taken to be achieved by a set of production processes by employing materials obtained from outputs themselves and human labor. Thus a productive process i is described as follows:

$$\text{Production process } i : (a_i, l_i) \Rightarrow b_i$$

Here a_i is a set of materials a_{ij} for process (or activity) i, l_i is a labor input indispensable for process i, and b_i is a set of outputs b_{ij} contributed from process (or activity) i.[16] $A=[a_{ij}]$ is an input matrix; $B=[b_{ij}]$ is an output matrix; $l=(l_i)$ is a labor vector. We will impose "the regularity condition" on our production system to guarantee the solutions, if desired.[17]

We must then have a price system as the valuation measure of price as well as a quantity system as the technically feasible activity measure:

$$p = (p_1, \ldots p_n)$$
$$q = (q_1, \ldots q_m)$$

Here p_j represents the price of commodity j; q_i is an activity level i. These measures then construct their evaluation systems. One is the price–cost equation system, the other the supply–demand equation system.

The price–cost equation is set up by the use of the uniform rate of interest r[18]:

$$[B - (1 + r)A]\frac{p}{w} \leq l \tag{1.25}$$

This means that sales minus input costs (interest payment included) is less than labor cost. We call this "price feasibility". Next, the supply–demand equation system is set up using a uniform rate of growth g:

$$q[B - (1 + g)A] \geq d \tag{1.26}$$

Here d_j represents a final demand of j consumed by labor. We call $d=(d_i)$ a final demand vector or a consumption vector. This means that supply minus intermediate input (growth factor added) is greater than final demand. We call this condition "quantity feasibility".

Feasibilities

Price feasibility: whether the price–cost equation of a subsystem is fulfilled with respect to nonnegative prices at a given rate of profit r.

Quantity feasibility: whether the demand–supply equation of a subsystem is fulfilled with respect to nonnegative activities at a given rate of growth g.

Thus there is a basic set for truncation of an initially given set of production processes under the rule of cost minimization and income maximization.

---

[16] a_{ij} represents the input of commodity j used in process i, as measured in the total output of commodity i. Similarly, b_{ij} represents the output of commodity j produced by process i.

[17] See Schefold (1976) and Schefold (1978), reprinted in Schefold (1997). See Chaps. 2 and 3 for details.

[18] The uniformity is easily transformed by taking into account "different rents".

Truncation rules

Cost Minimization: Min ql subject to $[B-(1+r)\mathrm{A}]p \leq wl$ Income Minimization: Min dp subject to $q[B-(1+r)\mathrm{A}] \geq d$

1.2.2.1 Wage Curve

An idea of real wages can be used to generate the factor price frontier. We distinguish the real wage ω from the nominal wage rate w. The real wage ω is a number of the unit basket of final goods d. The total spending of workers then is ωql, while the nominal total wage is the sum of wages for all workers, i.e. wql. According to our custom, we then adopt two kinds of normalization rules:

$$ql = 1; \quad w = 1. \tag{1.27}$$

It then holds:

$$\text{The wage curve: } \omega(r) = \frac{1}{dp(r)}. \tag{1.28}$$

The wage curve is also called the "factor price frontier" where the most efficient combinations of factor prices are represented. The factor prices here are the wage rate ω as the price of labor l and the rate of profit r (interest included) as the price of capital k.

Correspondingly, we have a similar idea based on the quantity frontier of the growth rate and the consumption rate (g, c). This is called the consumption curve.

$$c(g) = \frac{1}{q(g)l}. \tag{1.29}$$

1.2.2.2 The First Law of Economics

By the duality theorem of programming,[19] we know that at optimum:

$$ql = dp. \tag{1.30}$$

[19] See for example Nikaido (1968), a remarkable book on this issue.

Hence $c(g)$ is congruent to $\omega(r)$. The same argument applied to $\omega(r)$ also applies to $c(g)$. As argued in many literatures, in a general case of joint production, uniqueness of this equality is not necessarily maintained. If we should successfully define "entropy of production", it should be held that Production Q must be offset by Monetary Circuit C.

$$\oint \delta C = -\oint \delta Q. \tag{1.31}$$

This relationship is called the first law of economics in Mimkes (2010), confirmed by this equality coming from the duality theorem.

It then turns out from feasibilities and truncation rules as above that a chosen subsystem on the envelope depends on the set of (ω, r) and (d, g). More specifically, the net output will be generated by way of a choice of subsystems under the truncation rules. Even given the same net output there might be multiple subsystems to materialize it. So the actual net output will be determined after the specific path is known. In our modeling, this path may be known by the choice of distributive variable set (ω, r) in the case of value system.[20]

1.2.3 A Numerical Example of the von Neumann Model

Assume a case of three processes and two goods produced (Table 1.2).

Table 1.2 A numerical example a joint-production system (based on Schefold 1989, p.120 but slightly changed)

Input system $A = \begin{pmatrix} a_1 \\ a_2 \\ a_3 \end{pmatrix} = \begin{pmatrix} a_{11} & a_{12} \\ a_{21} & a_{22} \\ a_{31} & a_{32} \end{pmatrix} = \begin{pmatrix} 5.333 & 16 \\ 3.8 & 1.5 \\ 0.1 & 0.8 \end{pmatrix} \geq 0$

Output system $B = \begin{pmatrix} b_1 \\ b_2 \\ b_3 \end{pmatrix} = \begin{pmatrix} b_{11} & b_{12} \\ b_{21} & b_{22} \\ b_{31} & b_{32} \end{pmatrix} = \begin{pmatrix} 12.333 & 36 \\ 16 & 13.25 \\ 7.9 & 11.5 \end{pmatrix} \geq 0$

Labor input $l = \begin{pmatrix} l_1 \\ l_2 \\ l_3 \end{pmatrix} = \begin{pmatrix} 1 \\ 1 \\ 1 \end{pmatrix} \geq 0$

Final demand $d = (d_1, d_2) = (5, 6.5) \geq 0$

[20] In the case of the quantity system, this path may be known by the choice of (d, g).

1.2.3.1 The Derivation of Wage Curves

We then apply the rule of cost minimization to this system. This is a kind of nonlinear programming because the system contains a nonlinear term $(1+r)a_{ij}p$. The solution $p(r)$ therefore is expected to be erratic with respect to a variation of r. Our traditional solution for this problem is to use the method of "wage curves" of subsystems. Each wage curve is derived from each truncation of the whole system:

$$\omega^{\{i,j\}} = \frac{1}{dp^{\{i,j\}}(r)}. \tag{1.32}$$

Here superscript $\{i, j\}$ implies a subsystem of process i and j. $p^{\{i,j\}}$ then means the price system of subsystem $\{i, j\}$. In our example, we solve in this following manner:

Truncation rule

Step 0. Let the whole system be a system of two-commodity production by three technically feasible processes. Step 1. Find square subsystems. Step 2. Construct each wage curve on each truncated subsystem. Step 3. Construct the envelope from all wage curves.

At Step 1 we find nine ways for subsystems in the case of two-commodity production by three processes:

$$\binom{3}{1}\binom{2}{1} + \binom{3}{2}\binom{2}{2} = 9. \tag{1.33}$$

This breaks down into (i) six ways of a single process operation and (ii) three ways of two-commodity production by a two-process operation. So we can depict nine wage curves on the plane (w, r). The wage curves of truncated subsystems will be as follows (Fig. 1.6).

We can then make up the envelope of wage curves on the plane (w, r) (Fig. 1.7).

1.2.3.2 Complexity of Production

In our modeling, the size of input/output matrices and the number of truncations may indicate complexity. If we impose the regularity condition, there will be no truncation problem because we immediately obtain a solution without truncation. We suppose a general case. Given a production system $\{A, B, l\}$, we may truncate the whole system to get a subsystem. The number of subsystems will be

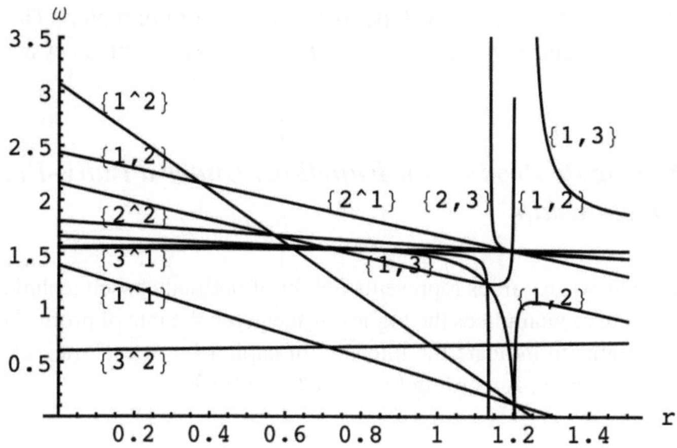

Fig. 1.6 Wage curves

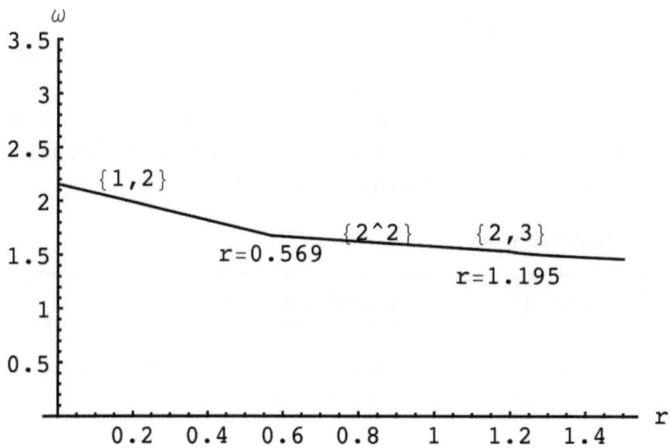

Fig. 1.7 The envelope of wage curves

$$\sum_{j=1}^{n}\binom{m}{j}\binom{n}{j}=\binom{m}{1}\binom{n}{1}+\binom{m}{2}\binom{n}{2}+\ldots+\binom{m}{n}\binom{n}{n}. \quad (1.34)$$

Here m is the number of processes, n is the number of commodities. It holds $m \geq n$. Let m be approximated by n, for simplicity. The total number of subsystems will be

$$-1+\frac{4^{M}\text{Gamma}\left[\frac{1}{2}+M\right]}{\sqrt{\pi}\text{Gamma}[1+\text{M}]}$$

But all of the large number of subsystems will not necessarily appear on the envelope, i.e. the efficient factor price frontier. Some of them will fail to be efficient. In other words, the envelope may lose some information. Thus we may require the idea of entropy to deal with such a huge number of subsystems.[21]

1.2.4 Surrogate Production Functions Under a Putty–Putty Relationship

The envelope of wage curves represents the set of optimal chosen techniques in the sense that the curve guarantees the highest wage given the rate of profit. The tangent of the curve seems to indicate the intensity of capital k[22] at each rate of profit r to produce a net output y. According to Samuelson (1962):

$$\begin{aligned} y^i &= \omega^i(r) + r \tan \gamma^i = \omega^i(r) + rk^i \\ \tan \gamma^i &= \frac{d\omega^i(r)}{dr} = k^i. \end{aligned} \tag{1.35}$$

Here i represents a subsystem employed, so y^i is regarded as the net output of subsystem i given the rate of profit r. The envelope of wage curves is called the "surrogate production function". Samuelson's hypothesis (1962) is based on a putty–putty relationship in production and income. He then predicted such a desired relationship that y^i is smoothly increasing with k^i:

Surrogate function hypothesis

Given y^i, k^i is uniquely determined, and vice versa.

If it should be true, we could safely solve an optimal subsystem any time. Hence we can enjoy the idea of aggregate capital k^i (Fig. 1.8).

1.2.5 Capital Controversies and the Path Dependency of the Choice of Technique

Unfortunately, the Samuelson prediction does not hold, as economists at Cambridge University (UK) pointed out. Just after the publication of Samuelson (1962) and Levahri and Samuelson (1966), many scholars observed so-called reswitching

[21] In the case of $m = 100$, the total number will be $9.05485 {*} 10^{58}$. But it is 135.753 after the log transformation of it.

[22] k: capital per capita or capital-labor ratio; d: consumption per capita; w: wage rate; r: rate of profit.

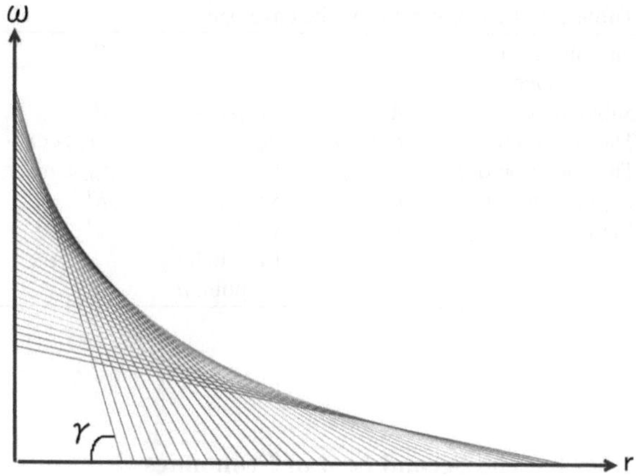

Fig. 1.8 A spectra of techniques

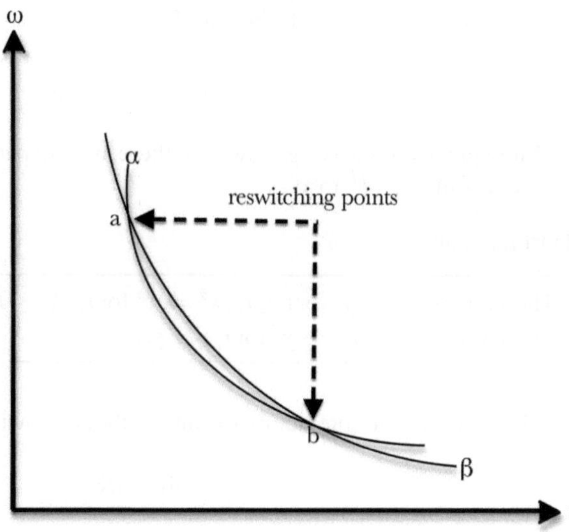

Fig. 1.9 A reswitching example

phenomena. The existence of complicated reactions between subsystems is easily verified. The envelope of wage curves must not remove the probability of emerging reswitching points. This phenomenon is called "Capital Reverse Deepening". Schefold (2008) illustrates this matter by introducing a recent simulation by Han and Schefold (2006): in the case of a Constant Elasticity Substitution (CES) production function, Capital Reverse Deepening may emerge by a very limited percent, e.g. 3.65%, while the remaining 96.35% are consistent with the traditional surrogate hypothesis (Fig. 1.9).

In this figure, the envelope of wage curves is to be arranged in the following manner:

Table 1.3 Different paths on the envelope

The surface of the envelope	A	$\Rightarrow$	B	$\Rightarrow$	α
Subsystem	A	$\{\alpha, \beta\}$	B	$\{\alpha, \beta\}$	α
The condition of ω	$\omega > \omega_a$	ω_a	$\omega_a > \omega > \omega_b$	ω_b	$\omega_b > \omega$
The condition of r	$r < r_a$	r_a	$r_a < r < r_b$	r_b	$r_b < r$
Capital intensity k	k^α	$K^{\{\alpha, \beta\}}$	K^β	$k^{\{\alpha, \beta\}}$	k^α
Net output y	y^α	$y^{\{\alpha, \beta\}}$	y^β	$y^{\{\alpha, \beta\}}$	y^α
		Reswitching point a		Reswitching point b	

1.2.5.1 The Second Law of Economics

Table 1.3 shows that the different conditions (ω, r) applied to the same set of production $\{A, B, l\}$ may lead to different results. Hence k^i is not uniquely determined by a given y^i. It then holds:

$$y^i \neq \omega^i(r) + rk^i(r) \tag{1.36}$$

More precisely speaking, however, there are two paths, namely, a path from k^α to y^α and a path from k^β to y^β:

Different paths

The path $\alpha : k^\alpha \Rightarrow y^\alpha$ for $r \leq r_a; k^\alpha \Rightarrow y^\alpha$ for $r_b \leq r \leq R$
The path $\beta : k^\alpha \neq k^\beta \Rightarrow y^\beta$ for $r_a \leq r \leq r_b$

Hence we cannot predict δy ex ante without knowing a particular k:

$$\delta y \neq dk. \tag{1.37}$$

This is rewritten by using an unknown multiplier λ:

$$\delta y = \lambda dk. \tag{1.38}$$

This is "the second law of economics" according to Mimkes (2010), although we still do not have a definite idea of entropy in our economic modeling. If we liked a neoclassical production function, we can apply entropy to it thanks to Mimkes (2010).

As shown in this case, the realization of a path depends on the specific condition (ω, r). In our economic modeling, the choice of techniques is assumed to be flexible ex ante. Each condition on k_i therefore leads to a different result by keeping a technological set flexible.

1.2.6 *Recycling of Production and Consumption Reconsidered*

Production may be depicted in association with two measures: complexity and temperature. In our modeling, each subsystem is composed of a combination of processes chosen from the production set. The number of combinatorials may measure the "complexity of production". Second, the height of welfare and consumption, the "temperature of production", measures the achievement of production.

We had a dual linear system of price and quantity based on the same set of production $In = \{A, l\}$. In this scheme the production process P will start from an input system $\{A, l\}$[23] and end up at an output system $Out = \{A, B, l\}$, and then the consumption process C will begin from Out ending up at In. The recycling of production thus is completed; the recycling circle is shaped between the input and output systems.

Due to this attribute of production, an output system based on In may be regarded as a more complex system because it contains a larger set to include the smaller set In. Thus the production path is a transformation from less complexity to more complexity. There is a minimal living standard even at the starting point of In.

As just argued, the production path may be described by the envelope of wage curves. Activation of a given production set depends on the deployment of the envelope. That is to say, economic conditions will regulate which inputs and outputs can be activated. After the specific path has been determined by such economic conditions as indicated by the wage curve and the like, the results will be confirmed. They will differ, though each result initially shares both the same input list and the same output list. That is to say, the technique such as the capital–labor ratio k^i to be employed will differ. In sum, a different activation will give a different path and a different complexity of production.

For purposes of illustration, let the horizontal axis measure the degree of ex ante complexity,[24] and the vertical axis measure the height of welfare. This completes Fig. 1.10.

1.2.6.1 A Remark on the Entropy Production Function

Mimkes (2010) successfully formulated the entropy production function based on a neoclassical production model of a one-way street. So far, however, we have discussed the so-called multi-sector model. Without the use of general coordinates, we could not decompose into a supposedly separate single operation as if it were a production function. Hence we must look for a surrogate idea of entropy in our economic modeling. This may be one of our most important tasks.

---

[23]Strictly speaking, inputs for production consist of $\{A, l\}$.

[24]Ex post the degree of complexity will differ for each path.

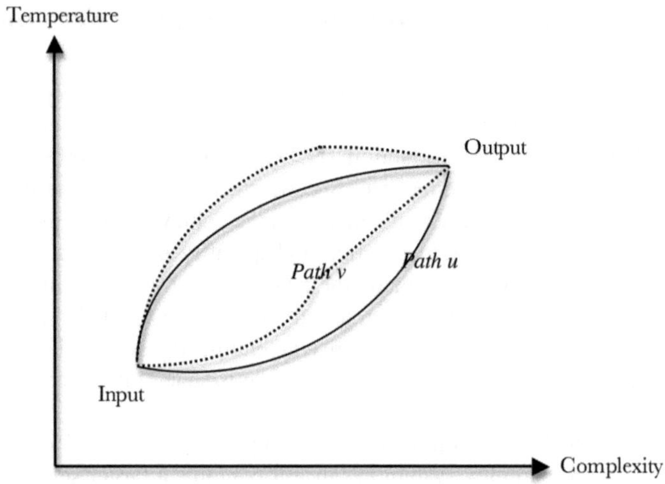

Fig. 1.10 An image of the second law on the choice of techniques

Finally, we refer to the second law in the entropy production function according to Mimkes (2010). The non-exact differential form δ:

$$\delta g(x, y) = a(x, y)dx + b(x, y)dy \tag{1.39}$$

is marked by δ.

$$\frac{\partial b}{\partial x} \neq \frac{\partial a}{\partial y} \neq \frac{\partial g}{\partial x \partial y}. \tag{1.40}$$

Integrals of non-exact differential forms are called Stokes integrals. They depend on the integral limits L and M and the path u of integration.

$$g_u(x, y) = \int_u \delta g(x, y). \tag{1.41}$$

The closed circle of δg along a circle on the plane (x, y) is not zero:

$$\oint \delta g(x, y) \neq 0. \tag{1.42}$$

The closed integral may be divided into two open reversible integrals between $L = (x_0, y_0)$ and $M = (x_1, y_1)$:

$$\oint \delta g(x, y) = \int_{x_0 y_0}^{x_1 y_1} \delta g + \int_{x_1 y_1}^{x_0 y_0} \delta g = \int_{x_0 y_0}^{x_1 y_1} \delta g - \int_{x_0 y_0}^{x_1 y_1} \delta g \neq 0. \tag{1.43}$$

The differential g, i.e.

$$g_u(x,y) = \int_u \delta g(x,y) \tag{1.44}$$

may be called an ex ante or putty differential form. The value g_u is putty, unpredictable, because every path u may lead to a different result and g may only be integrated after the specific path is known. The value of g_u is fixed ex post. The integrals along a closed line on the plane (x, y) will not cancel; they are putty or flexible. They have the same limits in the opposite direction, but the path u is different.

An unpredictable non-exact differential form (δy) may be expressed by an exact differential form (dg) and an integrating factor (λ). It then follows from the second law:

$$\delta y = \lambda dg. \tag{1.45}$$

The integrating factor λ may be defined in societies by mean income or standard of living, and in markets by the mean capital per participant in the market. The path u of integration has to be defined by a specified condition:

u: $\lambda = \lambda_0 =$ const. $\rightarrow Y_\lambda = \lambda_0 g$.
u: $g = g_0 =$ const. $\rightarrow Y_g = Y_0 =$ const.
u: Other conditions. Each condition leads to a different result by keeping the differential δy putty, flexible.

1.3 Human Interactions and von Stackelberg's Players: A Macroscopic Microeconomic Feedback

The slaving principle represents an integration of microscopic behaviours into a macroscopic order. In economics, there are several examples to introduce human interactions by taking into account a kind of macroscopic condition. Masanao Aoki, Hiroshi Yoshikawa, and others often noticed the importance of a "meso-macroeconomic method" (e.g. Aoki and Yoshikawa 2006). On the other hand, Hildenbrand suggested a "macroscopic microeconomic approach" (Hildenbrand 1994, p. 74). Hildenbrand (1994) verified that individualistic demand behaviour could be well-defined to make income effects always positive, provided that macroscopic variables like variances and covariances on spending among consumer goods are pertinently bound under a certain set of hypotheses. A microeconomic behaviour of demand could be solved by introducing a macroscopic condition. In this article, we will fix a perspective of a microeconomic game in view of macroscopic considerations by means of von Stackelberg's players.

We first have a brief look at traditional theories, in particular, game theory, and then establish the importance of heterogeneous interaction in a general framework, e.g. cluster dynamics with exchangeable agents. We then refer to a microeconomic interaction among two agents (firms) with exchangeable leadership by focusing the Stackelberg problem as Dynamic Bi-level Optimization. Finally, we take a microeconomic example to reformulate the equilibrium concept in a broader environment by taking into account positive feedback from some environmental changes.

1.3.1 Disengagements from the Efficient Market Hypothesis

Faced with a centennial economic crisis, Robert Lucas, Nobel laureate, in defending the dismal science (*The Economist* print edition, August 6, 2009), incredibly asserted that the current crisis even strengthened the credit of the efficient market hypothesis (EMH). As Ping Chen pointed out, however, the fundamental assumption behind the EMH is that financial markets are ruled by random walks or Brownian motion. If this theory were true, then it should be very unlikely to have large price movements like financial crises. Orthodox economics may no longer derive any prescription for this crisis from its own traditional theory.

1.3.1.1 Disengagement from the EMH

The current financial markets are filled with substantial secondary noise which could offset the primary efforts for regulation of demand and supply. The real financial markets are surrounded by too many non-market factors typical of the traditional stock exchange. The so-called sub-prime crisis essentially was irrelevant to the traditional stock exchange system, happening outside the stock market. Of course, the crisis heavily influenced the stock market. These factors thus amplify fluctuations. For example, in the currency exchange market, it is quite normal for professional traders to have mutual exchanges using private information, which seems like insider trading. But in the off-exchange trading of currencies, this is quite normal.

1.3.1.2 Economics of a Master Equation and Fluctuations

The real world economy actually seems to be irrelevant to the EMH. Interaction of heterogeneous factors inside and/or outside markets may generate many complicated outcomes in the world economy. This resembles the movements of exchangeable agents in a combinatorial stochastic process like the Urn process.

The stochastic evolution of the state vector can be described in terms of the master equation equivalent to the Chapman–Kolmogorov differential equation

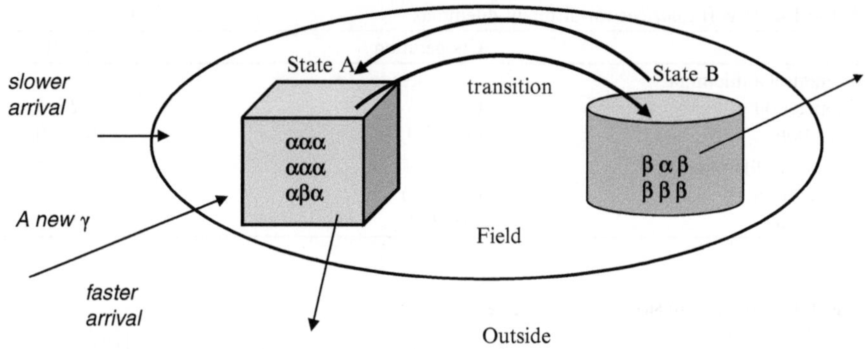

Fig. 1.11 An image of a master equation

system. The master equation leads to the aggregate dynamics, from which the Fokker–Planck equation can be derived. Thus we can explicitly argue the fluctuations in a dynamic system. These settings can be connected with the following key ideas making this feasible: classifying agents in the system by type and tracking the variations in cluster size. An intuitive image of such a situation will be depicted in Figs. 3.9 and 12.1 (Fig. 1.11).

1.3.1.3 Non-Self Averaging in Interactive Relationships

We take a two-person game with only positive spillover. In this game (Aruka 2001) selfishness may not be determined even if an agent selfishly adopts the strategy of defection. This game is called the "mutual fate control" game by the psychologists Thibaut and Kelley (1959). However, the expected value of gain for any agent could be reinforced if the average rate of cooperation were improved. This means we have a "macroscopically" weak control mechanism. Thus this could not always guarantee "mutual fate" directly.

Individual selfishness can only be realized if the other agent cooperates, therefore gain from defection can never be assured by defection alone. The sanction of defection by rival defection cannot necessarily reduce the selfishness of the rival. In this game, explicit direct reciprocity cannot be guaranteed. Aruka and Akiyama (2009) introduced different spillovers or payoff matrices, so that each agent may then be faced with a different payoff matrix.

Incidentally, the payoff structure of the Prisoners' Dilemma may be verified to be a slightly perturbed version of the Avatamsaka payoff if we employ a Tanimoto geometry (Tanimoto 2007a, b) (Table 1.4, Fig. 1.12).

Polya's Urn Process

A ball in the urn is interpreted as the number of cooperators, and the urn as a payoff matrix. Aruka and Akiyama (2009) apply Ewens' sampling formula to our urn process

Table 1.4 Payoff matrices for different dilemmas

	Cooperation	Defection
Avatamsaka dilemma		
Cooperation	(1, 1)	(0.25, 1)
Defection	(1, 0.25)	(0.25, 0.25)
Prisoners' dilemma		
Cooperation	(0.7, 0.7)	(0.25, 0.9)
Defection	(0.9, 0.25)	(0.3, 0.3)

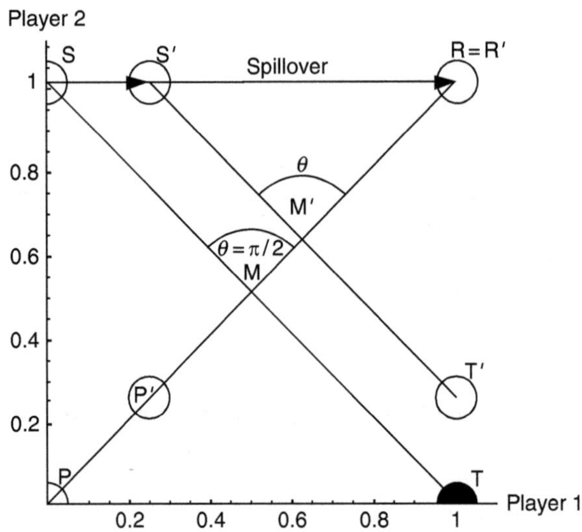

Fig. 1.12 An Avatamsaka payoff structure (See the same Fig.12.9.)

in this game theoretic environment. In this case, there is a similar result as in the classic case, because there is "self-averaging" for the variances of the number who cooperate. Applying Pitman's sampling formula to the urn process, the invariance of the random partition vectors under the properties of exchangeability and size-biased permutation does not hold in general.

Pitman's sampling formula depends on the two-parameter Poisson–Dirichlet distribution whose special case is just Ewens' formula. In the Ewens setting, only one probability α of a new entry matters. On the other hand, there is an additional probability θ of an unknown entry, as will be argued in the Pitman formula. More concretely, we will investigate the effects of different payoff sizes from playing a series of different games for newly emerging agents.

Increases in the Number of Cooperators and Non-Self Averaging

As Aoki and Yoshikawa (2006) dealt with a product innovation and a process innovation, they criticized Lucas's representative method and the idea that players face micro shocks drawn from the same unchanged probability distribution. In light of Aoki and Yoshikawa (2006), Aruka and Akiyama (2009), we show the same argument in our Avatamsaka game with different payoffs.

In this setting, innovations occurring in urns may be regarded as increases in the number of cooperators in urns whose payoffs are different.

1.3.2 Interactions in Traditional Game Theory and Their Problems

Even in traditional game theory, non-cooperative agents usually encounter various kinds of interactions in generating a set of complicated behaviours. In the course of finding curious results, experimental economists often are involved in giving game theorists new clues for solving the games. Their concerns are limited to informational structures outside the essential game structure, e.g. using such an auxiliary apparatus as "cheap talks". Seemingly, the traditional game theory is allowed to argue actual interactions extensively. But these attempts may have some difficulties, because the treatment of information structure is merely intuitive and not systemic.

1.3.2.1 The Limitation of Stochastic Independence of Mixed Strategies

We exemplify one limitation in the case of a usual Nash equilibrium. If we are faced with some multiple Nash equilibria, the concept of correlated equilibria may be activated. As Kono (2008, 2009) explored, however, this concept really requires the restrictive assumption that all players' mixed strategies are assumed to be stochastically independent. Without this requirement, a selection of equilibria may be inconsistent. Plentiful realistic devices for strengthening the traditional game theory cannot necessarily guarantee the assumption of stochastic independence of mixed strategies. So the traditional theory may often be compensated by evolutionary theory in order to argue realistic interactions comprehensively.

1.3.2.2 The Absence of Various Contexts/Stages When Strategies Are Chosen

At the very beginning of the Prisoners' Dilemma, psychologists detect the possibility of multiple (perhaps infinite) solutions in this game. They rightly know that humans depend largely on non-economic motives. In the original context, players' various psychic situations had been entirely excluded except for the utilitarian pay-off calculation common to all the players, i.e. any other code than Omertà could thus not contribute to preventing players from mutually defecting. The Prisoners' Dilemma can thus easily be reformulated into another stage, e.g. an evolutionary repeated game stage. It is nowadays trivial in the framework of the evolutionary game that a certain device of players' memories or hierarchical decisions could create cooperation by generating noise when strategies are chosen. However the explicit introduction of various contexts/stages may not necessarily be compatible with the individualistic rationality principle at any moment.

1.3.2.3 Heterogeneous Interactions in a General Framework

A general framework for encompassing various contexts/stages can be systematically proposed by focusing on the concept of information partition often used in statistical physics or population dynamics. It is easy to find such an application in the field of socio- and/or econo-physics. We can then argue that a player transitions into another player as the situation/stage changes.

The Exchangeable Agents as the Situation/Stage Changes

Here the exchangeable agents come out by the use of a random partition vector in the idea of statistical physics or population genetics. The partition vector provides us with the state information. We can thus argue the size-distribution of the components and their cluster dynamics with the exchangeable agents.

The Cluster Dynamics with the Exchangeable Agents

Define an at-most countable set, in which the probability density of transitions from state i to state j is given. In this setting, dynamics of the heterogeneous interacting agents give the field where an agent can become another agent. This way of thinking easily welcomes the unknown agents. This work was done mainly by Aoki and Yoshikawa (2006) when they were concerned with reconstructing macroeconomics.

1.3.3 von Stackelberg's Approach to Human Interactions as Dynamic Programming

We may further seek a kind of link to connect a general cluster dynamic with a particular micro-founded interaction. Specifically, we shall refer to a microeconomic interaction among two agents (firms) with exchangeable leadership. Heinrich von Stackelberg (1905–1946) remains an influential economist due to his masterpiece *Marktform und Gleichgewichit* (Market Structure and Equilibrium) (von Stackelberg 1934). His work is now appreciated outside economics, particularly in terms of dynamic programming.

1.3.3.1 The Stackelberg Problem as a Dynamic Bi-Level Optimization with Feedback Information

The reputation of Stackelberg's book attracts many contemporary readers, particularly those working in the field of duopolistic/oligopolistic competition. His work is appreciated for its precise depiction of the interaction of price/quantity leadership and its followers.

Unlike the Cournot game, in the duopolistic market, firms act sequentially, with the leader choosing a quantity first. The follower observes the leader's decision and chooses his quantity. At this moment, the decision making must be hierarchical. The hierarchical decision could furthermore be generalized from the fixed leadership to the alternating leadership with feedback information. This new reformulation is in line with a "Dynamic Bi-level Optimization" problem.

According to Nie et al. (2008b), this was first recognized by Chen and Cruz (1972) and Simaan and Cruz (1973), and subsequently studied by Basar and Olsder (1995), Li et al. (2002), Chen and Zadrozny (2002), Martin-Herran et.al (2006), Nie (2005, 2007, 2006, 2008a, b), Wie (2007) and Yang et al. (2007). As Nie et al. (2008a, b) noted this idea can be applied to the problems of "Tolls on the transportation network", "Stock corporations under Delaware Law", and "Contract bridge in a trump game".

Pu-yan Nie et al. (2008b, p. 537) mentioned the example of "tolls on the transportation network" in Labbé et al. (1999) as dynamic bi-level optimization with feedback information:

> We consider the revenues raised from tolls set on a transportation network. Assume there are two routes, A and B, between two places, and A is a highway. When traffic is seriously jammed on B, the great majority of drivers is willing to pass along A to save time if the tolls are not too high. At this stage, if the tolls are set too high, traffic will be affected negatively. On the other hand, low toll values will also yield low revenues. Thus, in this situation, one strikes the right balance by maximizing total revenues, subjected to the network users. This induces a two-level problem with the toll station playing the leading role at this stage. When B is not crowded, drivers can spend a little more time without the highway A. The drivers, in this situation, try to balance the time saved, subjected to the toll of the corresponding arcs. This also yields a two-level problem and the drivers now play the leading roles in decision making at this stage.

1.3.3.2 Extensive Forms of the Stackelberg Game

In the following diagram, we suppose that Player 1 has two alternative strategies (a, b), while Player 2 has three alternative strategies (α, β, γ). We may obtain the traditional game form if we replace "Player 2 observes the state and the action chosen by Player 1" with "Player 2 observes the state". In other words, Player 2 can decide irrespective of the opponent's decision in the traditional setting.

In Fig. 1.13, the Haurie diagram, Player 1 is depicted as the classical leader, constantly playing his role of the leader. Haurie called this diagram "sequential Stackelberg information structure".

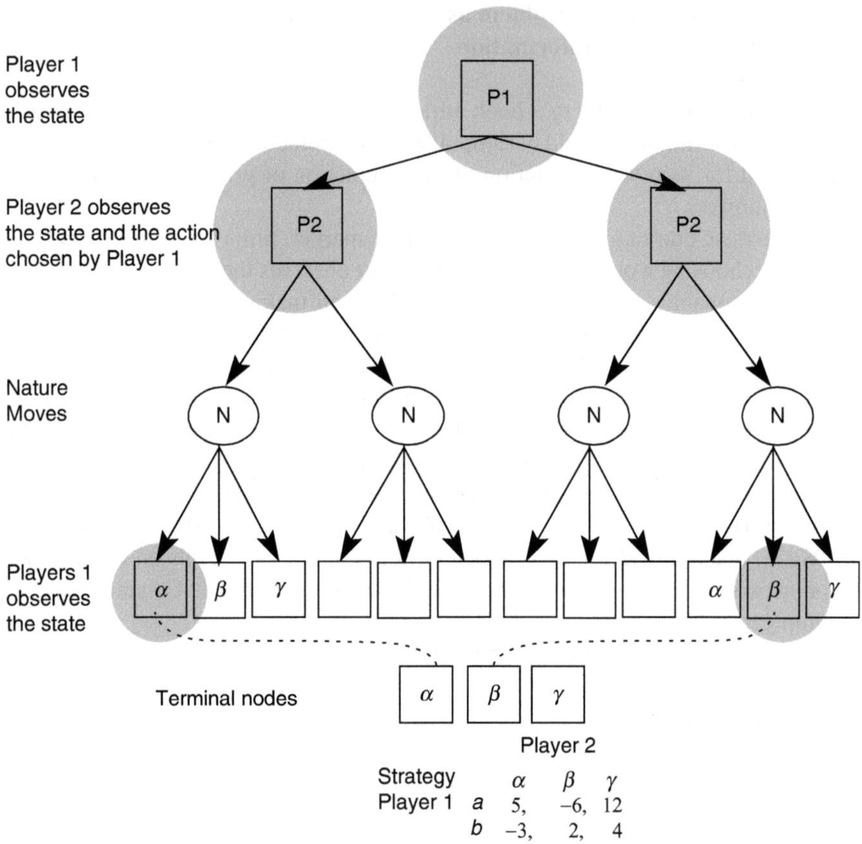

Fig. 1.13 The sequential Stackelberg information structure. Cited, with adding some modifications, from Figure 9.3: Haurie (1993)

The Alternating Exchange of Leadership Among the Players with Feedback Information

Due to Nie et al. (2008a, b), we may develop the diagram in which another evolution could be subsumed by contriving the part of "Nature moves". In short, "Nature moves" can be replaced with another institutional setting: the alternating exchange of leadership. The previous extensive form then loses a sequential information structure, and then the information structure degenerates at each stage due to the alternating exchange of leadership (see Fig. 1.14). We can no longer rely only upon the information structure if there is such a transitional change of leadership. The sequential decision process may be dominated by another dynamic principle.

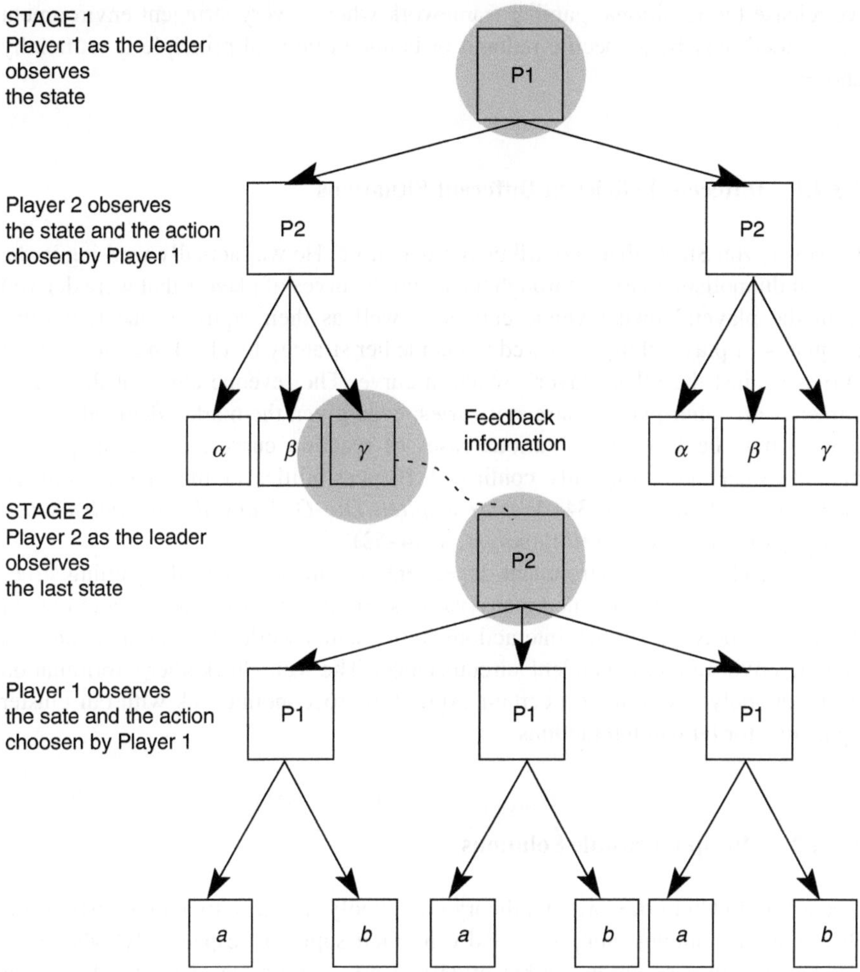

Fig. 1.14 The alternating exchange of leadership among the players with feedback information. This idea was proposed by Nie et al. (2008b).

1.3.4 von Stackelberg's Players' Reactions in Different Environments with Positive Feedback

Even in the case of the smallest game of two players with two strategies, there may be various kinds of solutions suggested. Sometimes the solution may be either a Pareto optimal or mini–max or Nash equilibrium. The Stackelberg introduction of players' leadership should change the game environment into a hierarchical decision structure, and the game may then be considered in terms of the bi-level optimization, i.e. a kind of dynamic programming. We can learn how we release the traditional gaming framework where a very stringent environment surrounded only by a specific rational or bounded rational principle is artificially chosen.

1.3.4.1 Different Policies in Different Situations

However, von Stackelberg contributed much more. He was actually working in the field of duopolistic markets through the reaction curves of players that were derived from the players' own revenue curves as well as their equi-revenue (isoprofit) isoquants. A player thus is allowed to decide her strategy by checking her isoquant curves against the other player's reaction curve. The revenue curve of the player varies as the other player's strategy varies, even given the market demand. So the players may be faced with various cases of reaction curves. In the duopolistic market, Stackelberg logically confirmed 16 possibilities of the market solution (see von Stackelberg (1934) *Viertes Kapitel. Die Ordnung der duopolistischen Marktpositionen* 1. *Generelle Analyse*, s. 44–53).

By introducing an environment-dependent circumstance into the gaming structure, he successfully imagined many various possibilities from the interactions of merely two players. Human interactions in the actual world also fluctuate due to a similar environment-dependent circumstance. The von Stackelberg formulation to the duopoly may hint at a certain explicit microeconomic link with our cluster dynamics for human interactions.

1.3.4.2 Multiple Feasible Solutions

According to Cournot's original theory on duopoly, we have two duopolistic firms that own their quelles. Suppose u and v are their supplies, respectively, when they are faced with a common market price p. Here is a simple example of a linear market demand curve:

$$p = 60 - X = 60 - (u + v). \tag{1.46}$$

In this setting, two firms obtain their profits pu, pv symetrically if their supply costs are set equally zero. Thus the reaction curves for the quantity strategy are as follows:

$$\frac{d\,pu}{du} = 60 - 2u - v = 0; \frac{d\,pv}{dv} = 60 - u - 2v = 0. \tag{1.47}$$

The solution for the above equation system is $u = v = 20$, assuring the profit of 400 for each firm. Around the solution $u = v = 20$ is a representative payoff matrix of Cournot duoply (Table 1.5, Figs. 1.15–1.17).

Table 1.5 A payoff matrix of Cournot duopoly

	$u = 20$	$v = 15$
$u = 20$	(400, 400)	(500, 375)
$v = 15$	(375, 500)	(450, 450)

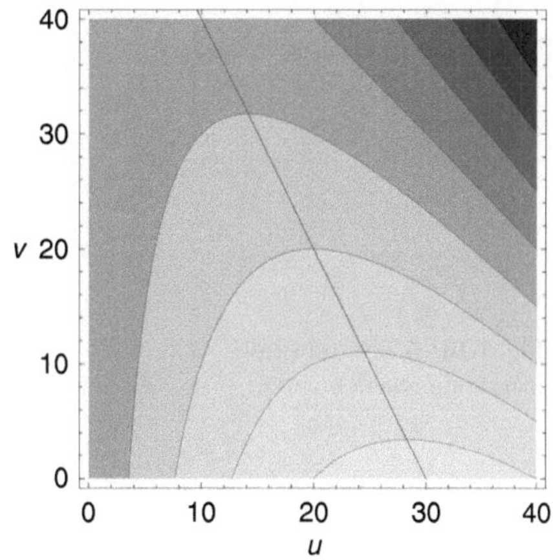

Fig. 1.15 The reactions line by u

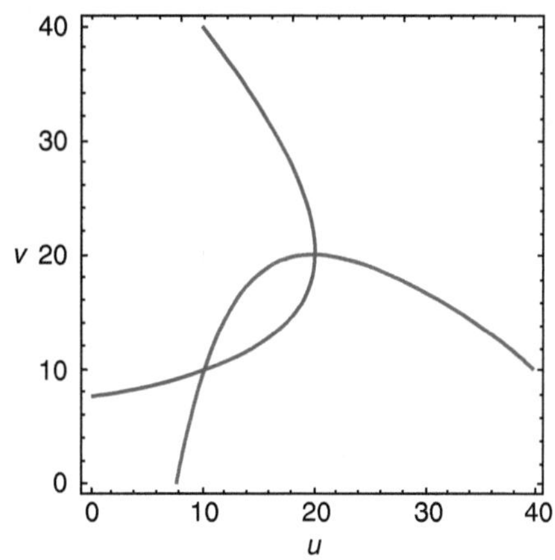

Fig. 1.16 The reactions line by v

Interestingly, using equi-revenue lines (from von Stackelberg), there is another solution that gives the same profit as 400. Thus this will cause a kind of multiple equilibria, and then a correlated equilibria as discussed in the Sect. 1.3.2.

As the environment changes, the equi-revenue curves will change. If we introduced a market demand curve with infinite elasticity of market demand with respect to price, there is an increasing reaction curve (Figs. 1.18 and 1.19).

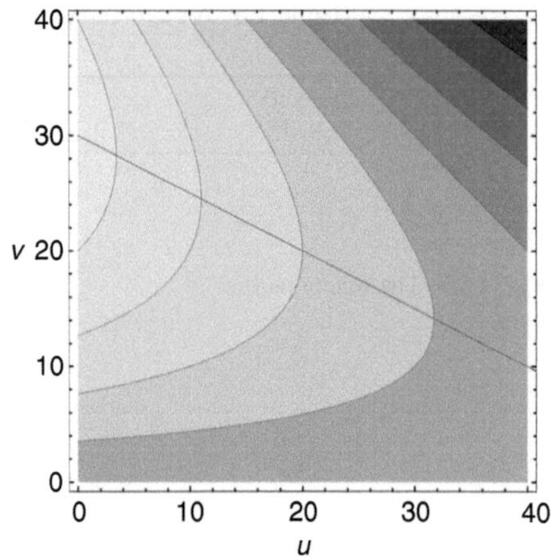

Fig. 1.17 Two crossing points of equi-revenue curves of 400. (u=v=10, p=40 leads to bringing the profit amount of 400)

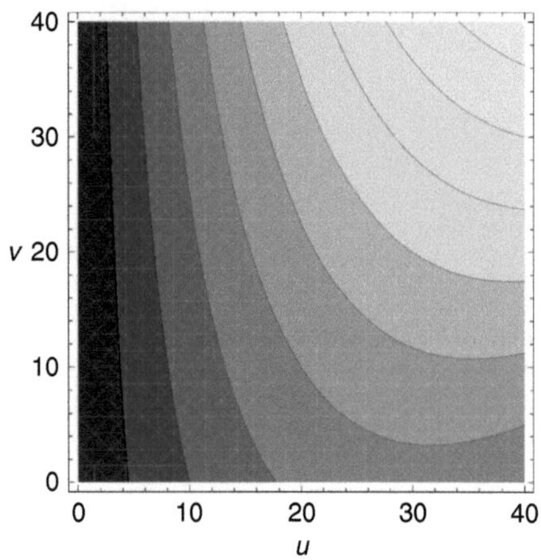

Fig. 1.18 A upwardward increasing reaction curve

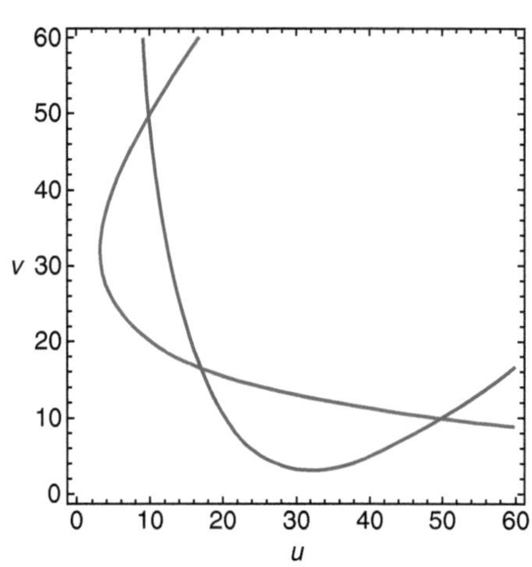

Fig. 1.19 Two crossing points of equi-revenue curves in the case of a upward increasing reaction curve

1.3.4.3 A Macroscopic Microeconomic Configuration

In von Stackelberg's view, the environment that a game is based in is not fixed to a specific environment. We can retrieve an evolution of game stages like alternating leaderships, taking into account different stages:

Agents: two symmetrical firms
Strategy: price or quantity
Reaction lines: upward or downward
Leadership: price or quantity

The combination of these different classes will give rise to $2\times2\times2\times2 = 16$ ways of equibrium. If we should add to the alternating rule, the number of equilibria would be increased. The notion of market equilibrium is then complicated. This may lead to a statistical and stochastic equilibrium in microeconomics.

References

Aoki M, Yoshikawa H (2006) Reconstructing macroeconomics: a perspective from statistical physics and combinatorial stochastic processes. Cambridge University Press, Cambridge, UK

Aruka Y (2001) Avatamsaka game structure and experiment on the web. In: Aruka Y (ed) Evolutionary controversies in economics. Springer, Tokyo, pp 115–132

Aruka Y, Akiyama E (2009) Non-self-averaging of a two-person game with only positive spillover: a new formulation of Avatamsaka's dilemma. J Econ Interact Coord 4(2):135–161

Barabási A-L, Albert A-L (1999) Emergence of scaling in random networks. Science 286:509–512

Basar T, Olsder GJ (1995) Dynamic noncooperative game theory, 2nd edn. Academic, New York, NY

Caon GM, Gonçalves S, Iglesias JR (2007) The unfair consequences of equal opportunities: comparing exchange models of wealth distribution. Eur J Phys Spl Top 143:69–74

Capocci A, Caldarelli G, De Los Rios P (2003) Quantitative description and modeling of real networks. Phys Rev E 68:047101

Chen CI, Cruz JB Jr (1972) Stackelberg solution for two person games with biased information patterns. IEEE Trans Autom Control 6:791–798

Chen B, Zadrozny PA (2002) An anticipative feedback solution for the infinite-horizon, linear-quadratic, dynamic, Stackelberg game. J Econ Dyn Control 26(9–10):1397–1416

Drăgulescu A, Yakovenkoa VM (2000) Evidence for the exponential distribution of income in the USA. Eur J Phys B 17:723–729

Drăgulescu A, Yakovenkoa VM (2001) Statistical mechanics of money. Eur J Phys B 20:585–589

Drouffe J-M, Godrèche C, Camia F (1998) A simple stochastic model for the dynamics of condensation. J Phys A 31:L19

Han Z, Schefold B (2006) An empirical investigation of paradoxes: reswitching and reverse capital deepening in capital theory. Camb J Econ 30(5):737–765

Haurie A (1993) From repeated to differential games: how time and uncertainty pervade the theory of games. In: Binmore K, Kirman A, Tani P (eds) Frontiers of game theory. MIT, Cambridge, MA, pp 165–193

Hildenbrand W (1994) Market demand. Princeton University Press, Princeton

Holland JH (1992) Adaptation in natural and artificial systems. MIT, Cambridge, MA

Holland JH (1995) Hidden order: how adaptation builds complexity. Basic Books, New York

Kono N (2008) Noncooperative game in cooperation: reformulation of correlated equilibria. Kyoto Econ Rev 77(2):107–125
Kono N (2009) Noncooperative game in cooperation: reformulation of correlated equilibria (II). Kyoto Econ Rev 78(1):1–18
Krapivsky PL, Redner S (2001) Organization of growing random networks. Phys Rev E 63:066123
Krapivsky PL, Redner S, Leyvraz F (2000) Connectivity of growing random networks. Phys Rev Lett 85:4629
Kuninaka H, Matsushita M (2009) Stastical properties of socio-physical complex systems. Kagaku, Iwanami, Tokyo 79(10):1146–1155
LabbÉ M, Marcotte P, Savard G (1999) A bilevel model of taxation and its application to optimal highway pricing. Manage Sci 44:1608–1622
Levahri D, Samuelson PA (1966) A reswitching theorem is false. Q J Econ 80:68–76
Li M, Cruz JB Jr, Simaan MA (2002) An approach to discrete-time incentive feedback Stackelberg games. IEEE Trans Syst Man Cybern A 32:472–481
Mainzer K (2007) (firstly published in 1994) Thinking in complexity. The computational dynamics of matter, mind, and mankind, 5th edn. Springer, New York
Martín-Herrán G, Cartigny P, Motte E, Tidball M (2006) Deforestation and foreign transfers: a Stackelberg differential game approach. Comput Oper Res 33(2):386–400
Mimkes J (2010) Stokes integral of economic growth: Calculus and the Solow model. Physica A 389:1665–1676
Newman MEJ (2002) Assortative mixing in networks. Phys Rev Lett 89:208701
Newman MEJ (2003) Mixing patterns in networks. Phys Rev E 67:026126
Nie PY (2005) Dynamic Stackelberg games under open-loop complete information. J Franklin Inst 342(7):737–748
Nie PY (2007) Discrete time dynamic multi-leader–follower games with stage-depending leaders under feedback information. Nonlinear Anal Hybrid Syst 1(4):548–559
Nie PY, Chen LH, Fukushima M (2006) Dynamic programming approach to discrete time dynamic feedback Stackelberg games with independent and dependent followers. Eur J Oper Res 169(1):310–328
Nie PY, Lai MY, Zhu SJ (2008a) Dynamic feedback Stackelberg games with non-unique solutions. Nonlinear Anal Theory Methods Appl 69(7):1904–1913
Nie PY, Lai MY, Zhu SJ (2008) Dynamic feedback Stackelberg games with alternating leaders. Nonlinear Anal Real World Appl 9:536–546 (also Corrected Proof, 2009, available)
Nikaido H (1968) Convex structures and economic theory. Academic, New York
Ohkubo J, Yasuda M (2005) Preferential urn model and nongrowing complex networks. Phys Rev E 72:0651041–0651044
Ohkubo J, Yasuda M, Tanaka K (2006) Replica analysis of preferential urn mode. J Phys Soc Jpn 75(7):1–6 [Erratum: 76(4), Article No. 048001 (2007, April)]
Pastor-Satorras R, Vázquez A, Vespignani A (2001) Dynamical and correlation properties of the Internet. Phys Rev Lett 87:258701
Samuelson PA (1962) Parable and realism in capital theory: the surrogate production function. Rev Econ Stud 29:193–206
Schefold B (1976) Relative prices as a function of the rate of profit. Zeitschrift für Nationalökonomie 36:21–48
Schefold B (1978) On counting equations. Zeitschrift für Nationalökonomie 38:253–285
Schefold B (1997) Normal prices, technical change and accumulation. Macmillan, London
Schefold B (2008) C.E.S. production functions in the light of the Cambridge critique. J Macroecon 30(2):783–797
Simaan MA, Cruz JB Jr (1973) A Stackelberg solution for games with many players. IEEE Trans Automat Contr 18:322–324
Tanimoto J (2007a) Promotion of cooperation by payoff noise in a 2 times 2 game. Phys Rev E 76:0411301–0411308

Tanimoto J (2007b) Does a tag system effectively support emerging cooperation. J Theor Biol 247:754–756

Thibaut JW, Kelley HH (1959) The social psychology of groups. Wiley, New York

von Neumann J (1937) Über ein Ökonomisches Gleichungssystem und eine Verallgemeinung des Brouwerschen Fixpunktsatzes, Ergebnisse eines Mathematischen Kollquiums 8 (1935-36), 73-83, Franz-Deuticcke, Leipniz aand Wien, 1937 (translated as A Model of General Economic Equilibrium. Rev Econ Stud 13(1945–46):1–9)

von Stackelberg H (1934) Marktform und Gleichgewichit (Market Structure and Equilibrium). Springer, Wien und Berlin (English translation 1952: H. von Stackelberg, The Theory of the Market Economy, Oxford University Press, Oxford, UK)

Wie BW (2007) Dynamic Stackelberg equilibrium congestion pricing. Transp Res C Emerg Technol 15(3):154–174

Wright I (2005) The social architecture of capitalism. Phys A Stat Mechanics Appl 346 (3–4):589–620

Xulvi-Brunet R, Sokolov IM (2005) Changing correlations in networks: assortativity and dissortativity. Acta Phys Pol B 36(5):1431–1455

Yang H, Zhang X, Meng Q (2007) Stackelberg games and multiple equilibrium behaviors on networks. Transp Res B Method 41(8):841–861

Part I
Complexities of Production and Social Interaction

Chapter 2
Generalized Goodwin's Theorems on General Coordinates

Yuji Aruka

This paper is concerned with providing foundations for the use of general coordinates in linear production systems, which Professor R. M. Goodwin originally proposed. The actually observed sectors of the Sraffa–Leontief system have bilateral relations to each other, which leads to complexities in analyzing prices or quantities. In the case that it is impossible to proceed with analysis directly in the original space, we may turn to an imaginary space, which is tractable, as long as the original space corresponds uniquely to the imaginary one adopted. This is a natural way for mathematics to proceed. Goodwin's use of the method of general coordinates represents a "diffeomorphisrn" in this area of economic theory. The use of general coordinates in the Sraffa–Leontief system seems, at first glance, restrictive, since their use requires a diagonalizable input matrix. A mathematical meaning of diagonalizability (or the rank condition) is examined by means of linear perturbation. The eigenvalues of the linear perturbed system are seen to be distinct almost everywhere. Economic meaning is given to the diagonalizable input matrix as yielding a regular system. It is worth noting that this specification can be justified in essentially the same way as the notion of quasi-smoothness used by Mas-Colell in the neoclassical production set. As a by-product of this study, the perturbed version of the Sraffa–Leontief price system is analysed.

2.1 Introduction

In order to make the Sraffa–Leontief system useful for dynamic analysis, Richard Goodwin (1983a) has recommended the use of eigenvectors of the square input matrix as "general coordinates". His recommendation has contained some

Reprinted from Structural Change and Economic Dynamics 2(1), Aruka, Y., Generalized Goodwin Theorems on General Coordinates, 69–91 (1991), http://www.sciencedirect.com/science/journal/0954349X. With kind permission from Elsevier, Oxford, UK.

Y. Aruka
Faculty of Commerce, Chuo University, Higashinakano, Hachioji, Tokyo 192–0393, Japan
e-mail: aruka@tamacc.chuo-u.ac.jp

DOI 10.1007/978-3-7908-2618-0_2, © Springer-Verlag Berlin Heidelberg 2011

overstatenents, as will be discussed in Sect. 2.2. The method of general coordinates, however, is no other than the use of normal modes in differential or difference equation systems. It will, under the assumption of the diagonalizability on the input matrix, be easily shown in Sect. 2.5 that the Sraffa–Leontief price system can be reformulated as a differential equation system in the rate of profit. The use of normal modes in the differential equation system decisively depends on the diagonalizability of the system. First of all, in Sect. 2.3, we shall point out that almost non-diagonalizable matrices are structurally unstable to perturbations (slight changes of coefficients of matrices), with the result that matrices after perturbation may be diagonalizable almost everywhere. Non-diagonalizable matrices are extremely vulnerable while diagonalizable matrices are structurally stable. By virtue of generalized general coordinates and perturbations, we can analyse a relation between a given unperturbed system and the perturbed system as argued in Sect. 2.4.

In Sect. 2.5 of this paper the diagonalizable Sraffa–Leontief system will be extended to the regular system. The diagonalizable Sraffa–Leontief system can be regarded as a standard for independent price movements in the extended system (the regular system). Anomalies of price movements will appear in the non-diagonalizable (irregular) system. Finally, in Sect. 2.6, we will call attention to features of perturbation effects in the diagonalizable Sraffa–Leontief price system, which may be a promising candidate for a new analysis of technological change in the linear multisectoral model.

2.2 Similarity Transformation and the Sraffa–Leontief System

The motion of two unequal masses in physics is a well-known application of the principals of transformation. In this example, we have two matrices A, B, where A is a symmetric matrix while B is not specified but called a mass matrix. Then we have the generalized eigenvalue problem:

$$Ax = \mu Bx. \tag{2.1}$$

The symmetric matrix A ensures a simultaneous diagonalization of the other matrix. The procedure of a simultaneous transformation due to a symmetric matrix is a "congruence transformation". It is occasionally called the principal transformation. Goodwin (1953) was successful in finding a would-be symmetric matrix in economics. This has been derived from the simultaneous adjustment system of price and quantity.

In this paper we will deal with the eigenvalue problem while dispensing with a symmetric matrix:

$$Ax = \mu x. \tag{2.2}$$

Here A is a square matrix. Diagonalization is of great advantage in this case. As is well known, in a differential equation system, a system of equations is decoupled by finding the eigenvectors:

"These eigenvectors are the 'normal modes' of the system, and they act independently. We can watch the behavior of each eigenvector separately, and combine these normal modes to find the solution. To say the same thing in another way. the underlying matrix has been diagonalized" (Strang 1980, p. 184)

The linear multisectoral model such as the Sraffa–Leontief system is reducible to the above form (2.2) of the eigenvalue problem. In this connection, Goodwin has recommended the use of eigenvectors of a square matrix as "general coordinates".[1] "On general coordinates, variables have been separated, analysis can proceed without concern for interdependence-without, however, leaving it out of account" (Goodwin and Punzo 1987, p. 51). Instead of the rectangular coordinates that are generally used, general coordinates are represented as oblique axes against the rectangular axes. In the linear production system dual to the linear price system, the transformation is also dual in the sense that the oblique axis of value is orthogonal to the dual output oblique axis, The whole system (either the price system or the quantity system) can be decomposed without resort to any variable.

Originally, Goodwin worked within the similarity transformation by assuming direct diagonalization, although square matrices are in general "defective matrices" whose eigenvectors are linearly dependent and, therefore, not diagonalizable. To begin with a well-known proposition for the similarity transformation with all distinct eigenvalues $\mu_1, \ldots, \mu_n$:

Proposition 1. *Suppose that the $n \times n$ matrix A has linearly independent eigenvectors. Then if these vectors are chosen to be the columns of a matrix H, it follows that $H^{-1}AH$ is a diagonal matrix $[\mu_{i\ diag}]$, with the eigenvalues $\mu_1, \ldots, \mu_n$ of A along its diagonal* (see, e.g. Strang 1980, p. 190, 5E).

In our supposition that the matrix A has no repeated eigenvalues, the eigenvectors are being chosen as a new basis for the space, since the n eigenvectors are automatically assumed independent. Thus H is constructed, and the underlying transformation represented by the diagonal matrix $H^{-1}A\,H = [\mu_{i\ \mathrm{diag}}]$ is obtained. In other words, a system of linear equations is decoupled by finding the eigenvectors. The following examples in the realm of Proposition 1 illustrate this (see Aruka 1987, p. 3):

Example 1. Take the eigenvectors as the oblique system (0; H), and then choose an imaginary price vector along H axes fulfilling

$$H\hat{p} = p,$$

where p is the actual price vector. Also define the linear transformation

[1]See Goodwin (1976, 1983b, c) and Goodwin and Punzo (1987). General coordinates are often called "principal coordinates" by Goodwin. See, for example, Goodwin and Punzo (1987, p. Chap. 2.5: Principal Coordinates).

$$p^* = (1 + r)\ Ap. \tag{2.3}$$

Here p^* is the actual price vector after the linear transformation, A is the non-negative input matrix and r may be interpreted as the uniform rate of profit, wage costs are ignored, However, take $\hat{p}^*$ along H giving $H\hat{p}^* = \hat{p}^*$, $\hat{p}^*$ is an imaginary price vector along H. Substituting $p = H\hat{p}$ into the above transformation equation system, $\hat{p}^*$is not only solvable but also completely decoupled[2]:

$\hat{p}_i^* = (1 + r)\mu_i\hat{p}_i$ for every eigensector i

$$\hat{p}^* = (1 + r)H^{-1}AH\hat{p} = (1 + r)[\mu_{i\ diag}]\hat{p}.$$

In this way, the imaginary price system along H in the oblique system proposed by Goodwin will be completely disaggregated, as if each sector were a Ricardian "Corn Economy".

In order to take account of the fixed cost in the price system, we may also choose an imaginary vector $\hat{k}$ along H, i.e. eigenrays of A:

$$H\hat{k} = k,$$

where k is an actual fixed constant cost vector. Applying the similarity transformation to the price system $[I{-}A]p = k$, we get

$[I - \mu_i]\hat{p}_i = \hat{k}_i$ for every eigensector i.

Here we should notice that subscripts i after the transformation do not indicate the actual sectors. The principal eigensector associated with $\mu(A) = \max\ \mu_i$, produces Sraffa's "Standard Commodity" in the dual subspace.

Example 2. Suppose the non-homogeneous difference equation system

$$q(t + 1) = (1 + r)^{-1}q(t)A + c(t), \tag{2.4}$$

when $q(t)$ is the non-negative output vector at period t, and $c(t)$ the autonomous demand at t. This economic system allows for different sectoral growth rates (see Goodwin 1976, 1983a, p. 143). The recursive solution of the system is given by

$$q(t) = (1 + r)^{-1}q(0)A^t + \sum_{k=0}^{t-1}(1 + r)^{-(t-k-1)}c(t)A^{t-k-1}.$$

Taking q along H^{-1} in line with Goodwin and substituting $A = H[\mu_{j\ diag}]H^{-1}$, the solution q is transformed into

[2]In the following, $[a_{i\ \mathrm{diag}}]$ indicates a diagonal matrix whose diagonal elements are a_i.

$$q(t) = (1+r)^{-1} q(0) \left[\mu_j^t{}_{diag}\right] + \sum_{k=0}^{t-1} (1+r)^{-(t-k-1)} c(k) \left[\mu_j^{t-k-1}{}_{diag}\right]. \tag{2.5}$$

This is the general solution expressed in terms of the eigenvectors of quantity H^{-1}. In more detail,

$$q(t)_j = (1+r)^{-1} \mu_j^t q(0)_j + \sum_{k=0}^{t-1} (1+r)^{-(t-k-1)} c(k)$$

for every j. If $q(0) = c(0)$,

$$q(t)_j = \sum_{k=0}^{t-1} (1+r)^{-(t-k-1)} c(k)_j.$$

Thus even the movements of the output system allowing unequal sectoral growth rates are decomposed and then combined. The whole movement is determined through this combination.

Example 2 in terms of the difference equation system corresponds precisely to the use of the notion of normal modes in the differential equation system. Example 1, however, also corresponds to the use of the normal modes. Goodwin has been confusing in this context. In Theorem 3 of Sect. 2.5, it will be easily shown that the Sraffa–Leontief price system can be reformulated as a differential equation system in the rate of profit. Prices of the Sraffa–Leontief system can then be regarded as a general solution in terms of a sum of separate eigenvectors (particular solutions). It will be verified that general coordinates of Goodwin are normal modes, each of which can be a definite constituent of the behaviour of a general solution. This is an essence of general coordinates. Each eigenvector corresponds to a particular solution. The whole movement of the system will not be explored without the general solution.

Goodwin himself has made some overstatements on how to use general coordinates which have vexed some economists. To quote Goodwin (Goodwin and Punzo 1987, p. 89):

"The theory of capital contains a logical contradiction: distribution depends on the quantity of capital, but the quantity of capital depends on distribution. Ricardo was aware of the problem and sought a way out by imagining a world of a single good, an unacceptable method which many since have had recourse to. The great advantage of the method of [general coordinates] is that it transforms the problem into a set of one-good problems without denying the existence of many goods. ... The numbers μ_1 to μ_n [eigenvalues of the input matrix] are invariant measures of capital intensity".

The final sentence, in italics in the above quotation, requires a more precise analysis. The input matrix A, if it satisfies the plausible assumptions such as non-negativity, irreducibility and producibility, guarantees that the principal eigenvalue $\mu(A)$ is real, simple, positive and smaller than 1. For the remaining eigenvalues, in which complex numbers may be contained, it follows that $|\mu|<\mu(A)<1$. In such a familiar setting, the role of the principal eigenvalue is illuminating. We mention one of the Frobenius Theorems applied to perturbation of essentially non-negative matrices:

Lemma 1. *$\mu(A)>\mu(B)$ if $A \geq B \geq 0$ (or $A - B \geq 0$) and at least either of them be irreducible.*[3]

It is easily seen from Lemma 1 that more inputs per unit of output from the original system A to a new system B imply a strict increase of the principal eigenvalue. The monotonicity of the principal eigenvalue in perturbation of a non-negative matrix is ensured as an invariant measure of capital intensity. However, the other eigenvalues will not be specified. Therefore Goodwin's incorrect exposition of this issue (Goodwin and Punzo 1987, pp. 91–92) has to be revised in the following manner[4]:

The principal eigenvalue $\mu(A)$ increases, [if] more inputs per unit of output are required, a capital-using change; [μ(A)] decreases, [if] it is capital-saving. The [principal eigenvalue] contains the essential dynamical structure of the whole square matrix representing the technology. Parentheses [ ] indicate corrections and [if] additions.

This illuminates a heuristic use of the Standard Commodity in a dynamical context, which few but Goodwin tried. The remaining eigenvalues and eigenvectors contribute to a general solution in Example 2.[5]

2.3 Diagonalization and Perturbation in the Linear Subspace

As long as Proposition 1 is valid, eigenvectors are the normal modes of the system, and they act independently. However, not all square matrices possess n linearly independent eigenvectors. Matrices are usually "defective matrices" in which eigenvectors fail to be bases. Whether we use a unitary or a Jordan canonical transformation on them, we only can at best triangularize any square matrix.[6]

---

[3]See, for example, Nikaido (1968, p. 103, Theorem 7.2 iv), Seneta (1981, pp 22–23). As for essentially non-negative matrices, see Kato (1982, p. 146. Theorem 7.2).

[4]In the original quotation, Goodwin considered each eigenvalue being eligible for a measure of capital intensity. He also confused a necessary condition for an increase of a principal eigenvalue with a sufficient condition for an increase of capital intensity. That is to say, he carelessly wrote: "If μ_i, increases, more inputs per unit are required".

[5][Sraffa] defines a standard commodity, which is the dominant eigenvector of his linear production system. But there are $n-1$ other output eigenvectors (and eigenvalues), and it proves illuminating to deploy all of them instead of merely the only purely real and positive one (Goodwin 1986, p. 205).

[6]See Strang (1980, p. 235) and also Aruka (1987, Sect. 11). "For any square matrix, there is a unitary matrix $S = U$ such that $U^{-1}TU = T$ is upper triangular" (Strang 1980, p. 234). This proposition is called "Schur's Lemma". If a unitary matrix satisfies that A Hermitian is equal to A Inverse, namely

From a practical point of view, it is nevertheless often argued that changing a few coefficients by infinitesimal amounts is enough to change it into a regular matrix.[7] Indeed, we know the following theorem:

Proposition 2. *Suppose that A is a matrix with repeated eigenvalues, or equivalently, A does not have a linearly independent set of n eigenvectors. Then there exists a matrix with distinct eigenvalues such that the norm*

$$\|A - B\| = \sum_{i=1}^{n}\sum_{j=1}^{n} \left|a_{ij} - b_{ij}\right| \leq \delta,$$

where δ is any positive number as small as is desired (see Klein 1973, p. 218; Gantmacher 1959, pp. 32–33).

We can write down $B = A + F$. We may regard F as an adjustment matrix to a given input matrix A. The version $A + F$ will then be considered as a case of constant coefficients in a broad sense where A contains small deviations or slight adjustments. Such a slight change is called a perturbation, We may use a small parameter$\varepsilon$to define a given matrix A under perturbation as $A(\varepsilon)$:

$$\begin{aligned} A(\varepsilon) &= A + \varepsilon A' \\ &= A + \varepsilon A^{(1)} + \varepsilon^2 A^{(2)} + \ldots \\ &= A + F(\varepsilon). \end{aligned}$$

In addition, we assume $F(\varepsilon) =\varepsilon F$. Then

$$A(\varepsilon) = A + \varepsilon F. \tag{2.6}$$

This version is a square matrix with linear perturbation where F is a perturbation direction matrix to specify the directions of perturbations, and ε is sufficiently small and may be complex. $A + \varepsilon F$ is occasionally called the perturbation matrix, while A is called the unperturbed matrix, εF the perturbation.[8] Proposition 2 is immediately stated in the context of linear perturbation.

In the following we state a number of propositions and theorems (proofs of which are in the Appendix) which prove the existence and examine the properties of the diagonalized form of the perturbed matrix $A + \varepsilon F$.

$A^H = A^{-1}$, the upper triangularization implies diagonalization. In the world of mutually adjoint (Hermitian conjugate) matrices, diagonalization is always guaranteed (Strang 1980).

[7]Blatt suggested an introduction of a slight adjustment to obtain desired result on the invertibility on the matrix. "[If] an irreducible matrix has no inverse, changing a few coefficients by infinitesimal amounts is enough to make it have an inverse; hence the conditions are not seriously restrictive" (Blatt 1983, p. 118).

[8]See e.g. Kan and Iri (1982, Chap. 5), Kato (1982. Chap. 2) and also Wilkinson (1965, Chap. 2).

Proposition 3. *Any defective square matrix A can be made to have all distinct eigenvalues by choosing an appropriate perturbation direction F for a sufficiently small $\varepsilon \neq 0$ (may be complex). Incidentally. diagonalization is valid for almost all perturbation directions F* (see Kan and Iri 1982, p. 177; Gantmacher 1959, pp. 32–33).

We continue to assume the linear perturbation $A + \varepsilon F$. Here F is an $n \times n$ matrix and a sufficiently small non-zero complex number. We can classify the effects of perturbation into two cases. One is the effect when we can arbitrarily choose the directions of perturbation. Then we virtually have $n^2 - 1$ degrees of freedom. The other is the effect under the case that an original configuration of algebraic multiplicity is kept after perturbation. In the latter case, the degrees of freedom in the directions of perturbation are restricted.

We now deal with two systems. One is the unperturbed system based on the matrix A, the other the perturbed system based on the matrix $A + \varepsilon F$. Algebraically, we have two polynomials $\det|A - \mu I|$, $\det|A - \varepsilon F - \mu(\varepsilon)I|$. The minimal polynomial of the latter is a polynomial in n^2 variables of εf_{kl}:

$$g(\mu) \equiv \mu^n + c_1\mu^{n-1} + \cdots + c_{n-1}\mu + c_n = c_0\Pi^n(\mu - \phi_i) = 0.$$

Here c, is at best an i-th order polynomial of εf_{kl}, while $c_0 = 1$. Let us define $a_0 = 1$, $a_i = c_i\ (i = 1, \ldots, n)$; $b_0 = n, b_j = (n-j)c_j\ (j = 1, \ldots, n-1)$. Then the resultant of $g(\mu)$ and $g'(\mu)$ is stated as follows:

$$\det\begin{vmatrix} a_0 & a_1 & \cdots & a_n & 0 & \cdots & 0 \\ 0 & a_0 & a_1 & \cdots & a_n & \cdots & 0 \\ \vdots & & & & & & \vdots \\ 0 & \cdots & & & a_0 & \cdots & a_n \\ b_0 & b_1 & \cdots & b_{n-1} & 0 & \cdots & 0 \\ 0 & b_0 & b_1 & \cdots & b_{n-1} & \cdots & 0 \\ \vdots & & & & & & \vdots \\ 0 & \cdots & & & b_0 & \cdots & b_{n-1} \end{vmatrix}.$$

Lemma 2. *$g(\mu)$ has multiple roots, if and only if the resultant of $g(\mu)$ and i.e., $R(g, g')$ is zero* (see e.g. Benavie 1972, pp. 162–163; Satake 1974, pp. 72–74).

Theorem 1. *Let A be any square matrix. The eigenvalues of $A + \varepsilon F$ are all distinct almost everywhere (see Kan and Iri 1982, pp. 117–178).* (For proof see Appendix.)

The next proposition immediately follows from the above inference.

Corollary 1. *If an unperturbed matrix A had no repeated roots, any perturbed matrix $A + \varepsilon F$ underlying A would have no repeated roots.* (For proof see Appendix.)

The diagonalizable matrix, in essence, is structurally stable. In Sect. 2.5, we will discuss the diagonalizable Sraffa–Leontief system and its extended regularities.

We introduce an eigenprojection P_k. Let $\Omega_{kj}(\varepsilon)$ be the generalized eigenvector space of $A(\varepsilon)$ associated with the eigenvalue μ_{kj}. $P_{kj}(\varepsilon)$ then is the subprojection operator onto $\Omega_{kj}(\varepsilon)$ along the remaining direct sum of $\Omega_{il}(\varepsilon)$ excluding $\Omega_{k1}(\varepsilon) + \cdots + \Omega_{kp_k}(\varepsilon)$:

$$Im\ P_{kj}(\varepsilon) = \Omega_{kj}(\varepsilon).$$

As ε converges to zero, $\mu_{kj}(\varepsilon)$ will converge to $\mu_k(k = 1, \ldots, s; j = 1, \ldots, P_k)$. If the unperturbed matrix A has multiple roots, we can find a set of the subprojections such that

$$Im\ P_{kj}(\varepsilon) = Im\ P_{kl}(\varepsilon)\ as\ \varepsilon \to 0.$$

It can be seen from the argument of Theorem 1 that the above overlapping images are regarded as measure zero. In this sense, the family of the diagonalizable matrix A (ε) implies a notion of regularity. It is worth noting that this specification can be justified in essentially the same way as the notion of quasi-smoothness used by Mas-Colell (1985) in the neoclassical production set.[9]

2.4 Economic Dynamics and the Use of Generalized Coordinates

We now analyse general coordinates without the diagonalization assumption. In order to demonstrate this relationship, we have to make use of such notions as eigenprojection and generalized eigenvector space. The eigenvector space spanned by the eigenvectors of a defective matrix A is called the generalized eigenvector space:

$$\left\{u_j \middle| [A - \mu_j I]^h u_j = 0\right\}.$$

In other words, we also have to add to the relation

$$Au_j = \mu_j \mu_j + u_{j-1}. \tag{2.7}$$

It is useful for us to illustrate this extension by means of a simple numerical example. (The example used is from Kan and Iri 1982, p. 166). Let A, F be as follows:

[9]"The production set Y is quasi-smooth if there is a closed region of Lehesgue measure zero Z such that the corresponding distance function C^2 is on $R^n \backslash Z$ (Mas-Colell 1985, p. 107, 3.6.1.)". As for the close similarity between our diagonality assumption and Mas-Colell's quasi-smoothness, see Aruka (1990).

$$A = \begin{bmatrix} 0 & 1 & 0 \\ 0 & 0 & 0 \\ 0 & 0 & 1 \end{bmatrix}, \ F = \begin{bmatrix} 0 & 0 & 0 \\ 1 & 0 & 0 \\ 1 & 0 & 0 \end{bmatrix}.$$

Suppose for instance A is an input matrix. We may then interpret A and F such that the second process (the second row) is non-existent (disguised) but that it will be realized after perturbation. The second process means a new activity industry. The perturbation will imply a "structural change" in a sense. The unperturbed system A has the repeated roots 0 (two zeros) and the simple root 1. The algebraic multiplicity of the root 0 does not coincide with geometric multiplicity. The eigenvectors associated with the eigenvalue 0 are [1,0,0]′ Thus we have to apply the above formula for the generalized eigenvector (2.7) and obtain the vector [0,0,0]′ The eigenvector associated with 1 is easily calculated as [0,0,I]′ (here ′ indicates transpose of vectors). In this case, we cannot construct the similarity transformation matrix H to diagonalize A. A Jordan cell of order 2 cannot vanish. On the other hand. $A + \varepsilon F$ for a sufficiently small $\varepsilon \neq 0$ has distinct eigenvalues $\sqrt{\varepsilon}, -\sqrt{\varepsilon}, 1$. The eigenvector accompanying $\sqrt{\varepsilon}$ is

$$u_{11}(\varepsilon) = [1, \sqrt{\varepsilon}, \varepsilon/(\sqrt{\varepsilon} - 1)]'.$$

The eigenvector accompanying $-\sqrt{\varepsilon}$ is

$$u_{12}(\varepsilon) = [1, -\sqrt{\varepsilon}, -\varepsilon/(\sqrt{\varepsilon} + 1)]'.$$

The eigenvector accompanying 1 is

$$u_2(\varepsilon) = [0, 0, 1]'.$$

As shown in Example 2 in Sect. 2.2, the solution of a difference equation system depends on its eigenvalues. For the sake of simplicity, we deal with the following homogeneous difference equation system under perturbation:

$$p(t+1) = [A + \varepsilon F]p(t) = A(\varepsilon)p(t). \tag{2.8}$$

The solution, if A (ε) is diagonalizable, is:

$$p(t) = H[\mu_{i\ diag}^{t}]H^{-1}u_0. \tag{2.9}$$

Here u_0 is the initial condition. We may regard $A + \varepsilon F$ as a perturbed input matrix, p as a price vector and the rate of profit as 0. We again specify $A + \varepsilon F$ in our numerical example. We can then solve the difference equation for $\varepsilon \neq 0$:

$$p(t) = H(\varepsilon)[\mu_{i\ diag}^{t}]H(\varepsilon)^{-1}u_0, \tag{2.10}$$

where

$$H(\varepsilon) = \begin{bmatrix} 1 & 1 & 0 \\ \sqrt{\varepsilon} & -\sqrt{\varepsilon} & 0 \\ \varepsilon/(\sqrt{\varepsilon}-1) & -\varepsilon/(\sqrt{\varepsilon}+1) & 1 \end{bmatrix}, \ [\mu_{i\ diag}^{t}] = \begin{bmatrix} (\sqrt{\varepsilon})^{t} & 0 & 0 \\ 0 & (-\sqrt{\varepsilon})^{t} & 0 \\ 0 & 0 & 1 \end{bmatrix}.$$

For $\varepsilon = 0$, the repeated eigenvalues 0 appear. However, the generalized eigenvectors corresponding to 1 are $[1,0,0]'$ and $[0,1,0]'$. The transformation matrix H happens to be the identity matrix I and its inverse exists. In the perturbation system we have all distinct eigenvalues $\mu_1 = \sqrt{\varepsilon}, \mu_2 = -\sqrt{\varepsilon}, \mu_3 = 1$, but the inverse of the transformation matrix $H(\varepsilon)$ is easily proved to be divergent for $\varepsilon \to 0$. Indeed, some components of $H(\varepsilon)^{-1}$ contain a factor $1/(\mu_1 - \mu_2)$. Suppose μ_1 and μ_2 are too close to be discernible: $\mu_1 \approx \mu_2$. In this case the solution will be unstable for the perturbation of order ε. However, the eigenvalues which are almost the same excepting the difference of order $\varepsilon^{1/r}$ might be considered a result of an error in estimates. Instead of the two distinct eigenvalues, we adopt an average of these as a surrogate value:

$$\mu_0 = (\mu_1 + \mu_2)/2 = (\sqrt{\varepsilon} - \sqrt{\varepsilon})/2 = 0.$$

These are just the repeated eigenvalues 0 in the unperturbed system. In general, the average of the close eigenvalues $\sum_{j=1}^{m} \mu_j/m$ does not always coincide with the repeated eigenvalues in the unperturbed system. Rather we demonstrate that the multiplicity of the eigenvalues in the unperturbed system is preserved in the perturbed system as a choice of the weighted mean.[10] We note that such a perturbation to preserve the multiplicities in the unperturbed system would increase the rank of its perturbation matrix.

Conversely, let the distance of the eigenvalues be $|\mu_1 - \mu_2| \gg \varepsilon^{1/2}$. It is easily verified from the inference on the convergence velocity of the eigenvalues and vectors that a change of the solution cannot respond much to a perturbation of order ε. In fact, instability of the solution will mainly depend on the order of the distances of the mutual eigenvalues $|\mu_j - \mu_k|$. These considerations are valid for a general case (see Kan and Iri 1982, pp. 185–190). Thus we could conclude as follows:

Theorem 2. *Given perturbation, the solution of the difference equation system $p(t+1) = [A + \varepsilon F]p(t)$ will not be greatly changed if the distances of the mutual eigenvalues $|\mu_j(\varepsilon) - \mu_k(\varepsilon)| \gg |\varepsilon^{1/r}|$. It will be divergent or oscillate if $|\mu_j(\varepsilon) - \mu_k(\varepsilon)| \leq |\varepsilon^{1/r}|$ where the equality implies a near equality.*

[10]The weighted mean of the μ-group eigenvalues of $A(\varepsilon)$ with the multiplicity h is

$$(1/h)\text{trace}[A(\varepsilon)P(\varepsilon)] = \mu + (1/h)\text{trace}\{[A(\varepsilon) - \mu I]P(\varepsilon)]\}.$$

According to this theorem, we can use general coordinates even under perturbation, unless all two of the eigenvalues of the perturbation system are mutually too close. Theoretically, we should note that we could not predict the perturbation effects "with certainty" when we were faced with such a perturbation as $\left|\mu_j(\varepsilon) - \mu_k(\varepsilon)\right| \le \left|\varepsilon^{1/r}\right|$.[11] Otherwise, particularly when the unperturbed system is defective, we have to retreat toward a defective (not diagonalizable) system, whose rank as usual is greater than the rank in the unperturbed system, by manipulating an average eigenvalue in order to assure a plausible solution, The latter treatment depends on practical judgement. In that case we have to leave general coordinates in the original sense given by Goodwin, We can nevertheless suggest the use of the generalized eigenvectors. Goodwin called the use of the eigenvector space "general coordinates". We call the use of generalized eigenvectors "generalized general coordinates".

Finally, we comment on an application of generalized general coordinates although in a particular case. We call the formula for the generalized eigenvector (2.7):

$$Au_j = \mu_j u_j + u_{j-1}.$$

In general, we can construct the Jordan form for any defective matrix if the transformation matrix supplemented with the generalized eigenvectors K are adopted. In our example, K happens to be the identity matrix but the Jordan form is rightly expressed. The diagonals 0, 0, 1 correspond to the eigenvalues and the off-diagonal 1 appears on the block of the repeated eigenvalues. Set $\mu_1 = 0$, $\mu_2 = 1$. Then our Jordan form is

$$J = K^{-1}AK = \begin{bmatrix} \mu_1 & 1 & 0 \\ 0 & \mu_1 & 0 \\ 0 & 0 & \mu_2 \end{bmatrix}.$$

We reproduce the following simple homogeneous price system underlying the matrix A:

$$p = (1+r)Ap.$$

This system originally lacks the real second process, as pointed out at the beginning of this section. Here we adopt "generalized general coordinates". Taking $\hat{p}$ along K, it follows that

$$\hat{p} = (1+r)\,J\hat{p}.$$

[11] A coefficient system when being faced with perturbation seems inherently unstable. Aruka (1988) discussed the uncertainty principle in the simultaneous adjustment system of price and quantity underlying the Sraffa–Leonticf system.

Writing this down in detail,

$$\hat{p}_1 = \mu_1\hat{p}_1 + \hat{p}_2, \hat{p}_2 = \mu_1\hat{p}_2, \hat{p}_3 = \mu_2\hat{p}_3.$$

The first equation cannot be separated from the second generalized eigensector. However, the number of the actual sectors is two in our numerical example. We only need two eigensectors in general coordinates before perturbation. In terms of general coordinates, we may regard the second generalized eigensector as a redundant one. In this case, we can in essence return to the world of general coordinates and accomplish a complete decoupling.

2.5 The Diagonalizable Sraffa–Leontief Price System

In this section, we reformulate the Sraffa–Leontief price system as a differential equation system in the rate of profit. Suppose a linear production system without joint production, where each process produces a single product. Labour is the sole primary factor of production and is indispensable in operating each process. A is an $n \times n$ square matrix of input coefficients a_{ij}, a^0 a labor input vector, p a price vector and r the rate of profit. The simplest price system will then be an in-homogeneous equation system:

$$[I - (1 + r)A]p = a^0. \tag{2.11}$$

Here the price of labor, the wage rate, is fixed at unity and prices of produced goods are therefore expressed in terms of labor. As argued by many authors, we could get as solution a non-negative vector by virtue of the following assumptions:

(1) An input matrix A is productive: $q\ [I - A] \geq 0$ for any $q \geq 0$.
(2) An input matrix A as a whole is non-negative and irreducible and its eigenvalue is smaller than $1/(1 + r)$.

Either of them guarantees a non-negative solution:

$$p(r) = [I - (1 + r)A]^{-1}a^0. \tag{2.12}$$

In particular, (2) confirms a unique non-negative solution. We now prove the transformation of the Sraffa–Leontief price system in view of the following lemma on commutability of matrices:

Lemma 3. [12]*Matrices with the same eigenvectors must commute. Its converse is also true.*

[12]See e.g. Strang (1980, p. 193). It is also noted that if $AB = BA$ and both A and B can be diagonalized, then they can be diagonalized by the same H.

Theorem 3. *Suppose the input matrix A is diagonalizable, non-negative and irreducible. Also assume that the principal real eigenvalue μ_1 is smaller than 1/(1 + r). We transform the Sraffa–Leontief price system*

$$p = (1 + r)Ap + a^0 \tag{2.13}$$

into the linear differential equation system in the rate of profit r:

$$\frac{dp_i}{dr} = \sum_{j=1}^{n} a_{ij}p_j + (1 + r)\sum_{j=1}^{n} a_{ij}\frac{dp_j}{dr} \text{ for } i \neq j. \tag{2.14}$$

(i) *Then the general solution of this system is*

$$p(r) = \sum_{i=1}^{M} e^{m_i r} c_i u_i \tag{2.15}$$

where $m_i \in M, M = [(1/\mu_i) - (1 + r)_{diag}]^{-1}, c = H^{-1}$ and H is a matrix of similarity transformation to A, H consists of the columns of eigenvectors u_i of A.

(ii) *Then*

$$Ap = \frac{m}{1 + m(1 + r)}p. \tag{2.16}$$

Incidentally, the eigenvalues of the differential system correspond to the eigenvalues of the Sraffa–Leontief price system by the relationship of $\mu/(1 - \mu(1 + r))$. (See proof in Appendix.)

It is seen that the relative prices are unchanged for any rate of profit r in terms of the i-th particular solution:

$$\frac{p_j}{p_1} = \frac{u_{ij}}{u_{i1}} \text{ for any } r. \tag{2.17}$$

Therefore, as long as the principal non-negative real eigenvector is adopted as a particular solution, the constancy of non-negative relative prices is assured whatever the rate of profit is. This is another meaning of the principal eigenvector. However, the actual prices have been represented as the general solution, which is a sum of particular solutions $\sum_{i}^{M} e^{m_i r} c_i u_i$. Then, the relative prices could no longer be unchanged as the rate of profit changes.[13]

Additionally, we mention two corollaries of Theorem 3(*i*):

Corollary 2. *In the differential equation system of Theorem 3, the general solution will be divergent, as* $1/(1+r) \to \mu(A)$. *Here* $\mu(A)$ *is the principal eigenvalue.* (Proof in Appendix.)

Corollary 3. *Let the assumptions of Theorem 3 hold. If the labor input* a^0 *should be a right-hand eigenvector of matrix A,* $p(r)$ *can be expressed in terms of* a^0 *only, other eigenvectors being neglected.* (Proof in Appendix.)

It is obvious from the corollary that such an anomaly as $Aa^0 = \mu a^0$ leads to a labor theory of value. The economic meaning of $Aa^0 = \mu a^0$ is the assumption of an equal organic composition of capital. Trivially, it also follows that if $sa^0 \neq 0$ for any left-hand eigenvector s of A, $Aa^0 \neq \mu a^0$. However, if $sp = 0$ for any right-hand eigenvector p of A, the quantity (activity) vector s is estimated as an eigenvector. In particular, s must be the Frobenius vector because an actual vector is non-negative. In this case, a particular solution may be regarded as a general solution. Relative prices are unchanged in each particular solution. In an economy producing the Frobenius vector s, therefore, relative prices also are unchanged against any change of distributive shares. This is another meaning of Sraffa's standard commodity. We can now prove Theorem 3 alternatively. The next theorem is equivalent to Theorem 3.

Theorem 4.[14] (i) *Suppose* $sa^0 \neq 0$ *for* $s \neq 0$, *any left-hand eigenvector of matrix* A *(or* $a^0 \neq (1+R)Aa^0$*). If and* $a^0, Aa^0, A^2a^0, \ldots, A^{n-1}a^0$ *are linearly independent, does there then exist a p such that*

$$Ap = \frac{m}{1+m(1+r)} p \text{ for some non-zero number } m.$$

(ii) *Normal modes of the Sraffa–Leontief price system are equivalent to the necessary and sufficient condition (i) of this theorem.* (Proof in Appendix.)

---

[13] Goodwin failed to formulate an analytically transformed system such as (2.14), with which eigenvectors are to be integrated as a general solution. However, each particular solution $e^{mr}u$ is connected with each other through a common rate of profit r. We replace (2.13) with the next one where wages are paid in advance:

$p = (1+r)(Ap+a^0)$.

Transforming this system onto general coordinates, it follows

$\upsilon_i = (1+r)(\mu_i \upsilon_i + b_i)$.

Here $p = Hu$, $a^0 = Hb$. Thus

$\mu_i + b_i(1/\upsilon_i) = 1/(1+r)$.

This is the relationship that Goodwin preferred to use (see Goodwin, 1983a, pp. 157–158; Goodwin-Punzo, 1987. p. 92).

[14] Similar statements of this theorem (i) were originally discussed in Schefold (1976) and Miyao (1977). However, they did not state the equivalence of this theorem to Theorem 1.

By Theorem 4, the assumptions of Theorem 3 have been revealed. An economic meaning of Theorem 4 is straightforward. The uniqueness of z of (2.31) depends on the rank of matrix L. Hence an independent price movement is confirmed if L has a full rank. The latter condition is a relation of A and a^0 in our linear single production system $[A, I, a^0]$. This relation corresponds precisely to the condition of equation system (2.14) which describes the movements of prices being equipped with "normal modes" and the initial condition determined by a^0. In particular, non-singularity, irreducibility, diagonality of matrix A together may be regarded as linear independency of matrix L and $sa^0 \neq 0$. We can therefore state the definition of regularity and irregularity in our linear production system. The distinction will be made whether a system is reduced to normal modes or not:

Definition 1. *(Schefold 1978, p. 267). The linear production system $[A, I, a^0]$ is called "regular" if the following conditions are fulfilled:*

(i) A is non-singular, i.e. $\det|A| \neq 0$.
(ii) $\mathrm{rank}[I-(1+R)A] = n-1$ (rsp. $\mathrm{rank}[A-\mu I] = n-1$.
(iii) $sa^0 \neq 0$ for any left-hand eigenvector s of $(1+R)sA = s$.
(iv) $qu \neq 0$ for any right-hand eigenvector u of $(1+R)Au = u$.

Here it is noticed that an eigenvalue μ is replaced with $1/(1+R)$.

Condition (i) implies that any pure consumption goods not being available in production are not contained in our system. By (ii), A is diagonalizable; (iii) excludes such a peculiar case as a^0 happening to be an eigenvector of A, while (iv) assures that our system, by accident, does not produce an eigenvector such as the Sraffa standard commodity. If $gu = 0$, it follows for at least one i that

$$u_i(r) = \frac{1}{g_i}(g_1u_1 + \cdots + g_{i-1}u_{i-1} + g_{i+1}u_{i+1} + \cdots g_nu_n), \tag{2.18}$$

where $g_i \neq 0$. This implies that a price of commodity i is a linear function of the remaining commodities produced. In this sense, (iv) guarantees "an independent price movement" for each commodity. Conditions (iii) and (iv) may be regarded as anomalies without reservations (see Miyao 1977, p. 161; Schefold 1976, p. 29).

2.6 Some Remarks on the Perturbed Version of the Sraffa–Leontief Price System

The constancy of technology is a problematic notion in economics. There were some attempts to relax the rigidity of linearity in the multisectoral model. Efforts were almost always directed to non-linearize a matrix A or to relate A to a scale factor q. These kinds of trials concern the solvability of the equation system:

$$q[A(q) - \mu I] = 0.$$

The existence of the solution is demonstrated by using Jacobian matrices. Alternatively, our perturbed version of the Sraffa–Leontief system mainly concerns the solvability of the equation system:

$$q(\varepsilon)[A(\varepsilon) - \mu(\varepsilon)I] = 0 \text{ for } \varepsilon \neq 0.$$

The actual input substitution reflects the perturbed version of the linear input system. The probable input substitution cannot be of the kind such as one between steerings and brakes, but rather the one among steerings or brakes. In general, it is not within producers' discretion that they switch between existing techniques as much as they desire. Possible short-run alterations to the process of production are normally limited to small deviations from a system originally adapted by producers. Therefore a small correction of their coefficients in an input matrix seems a promising candidate, since it will not give rise to grave adjustment problems. This is the reason why we adopted the perturbed version. Incidentally, it is important to point out that our perturbed version of the Sraffa–Leontief system resembles the C.E.S. production function (see Blatt 1983, pp. 119–148).

In general, the perturbed version of the Sraffa–Leontief price system will be of the form:

$$[I - (1+r)(A + \delta A)](p + \delta p) = a^0 + \delta a^0. \tag{2.19}$$

By subtraction,

$$\delta p - (1+r)\delta A \delta p = \delta a^0.$$

Hence we obtain the resulting solution:

$$\delta p = [I - (1+r)\delta A]^{-1}\delta a^0. \tag{2.20}$$

If a matrix $[I - (1+r)\delta A]$ has a dominant diagonal for a given r, there will exist an inverse matrix. In this particular case, we need not have a non-negative matrix since directions of perturbation are arbitrary. We know that a matrix has a dominant diagonal if a matrix has a quasi-dominant diagonal (McKenzie 1960, p. 48). It is, however, difficult to find an economically meaningful reason by which δA has a quasi-dominant diagonal. If we should find it, such a quasi-dominant "diagonality would depend on the rate of profit r". It is therefore a difficult problem for us to estimate price effects due to perturbation, although we can mention a few results of price effects in some special types of perturbation. At first, if only a labor input changes while the other inputs are constant, namely $\delta A = 0$, we obtain a simple interpretation:

$$\delta p = [I - (1+r)A]^{-1}\delta a^0. \tag{2.21}$$

Indeed, δp is definite as far as the matrix in the unperturbed system $[I - (1 + r)A]$ has a dominant diagonal for a given r, which means

$$\frac{1}{1+r} > \mu(A). \tag{2.22}$$

Here $\mu(A)$ is the principal (dominant) eigenvalue of the non-negative matrix A. Secondly, if, on the contrary, A changes while labor input remains constant, a weaker result follows. Since

$$\delta p = [I - (1+r)A]^{-1}[(1+r)\delta A](p + \delta p),$$

we have the following inequality making use of the norm of the matrix:

$$\frac{\|\delta p\|}{\|p + \delta p\|} \leq c \frac{\|(1+r)\delta A\|}{\|I - (1+r)A\|}. \tag{2.23}$$

Here $c = \|[I - (1+r)A]^{-1}\| / \|I - (1+r)A\|$. c is called the condition number of the matrix $[I - (1 + r) A]$.[15] By means of the non-negative irreducibility of the matrix A we can derive some definite results of the price effects after perturbation.

Proposition 4. *(Dietzenbacher 1988, p. 391). Let $\widetilde{\Gamma}$ be the sets where the indices {l} of the perturbed process are contained, while $\widetilde{\Gamma}_0$ has its complement set {k}. Also define $A, \widetilde{A}$ as the unperturbed matrix and the perturbed matrix, respectively. We assume the unperturbed matrix is non-negative and irreducible but the matrix after perturbation is still irreducible. Then*

$$\frac{\widetilde{p}_k}{p_k} \leq \frac{\mu(A)}{\widetilde{\mu}(\widetilde{A})} max_{j \in \Gamma} \frac{\widetilde{p}_j}{p_j} \text{ for } \mu(A) > \frac{\mu(A)}{\widetilde{\mu}(\widetilde{A})}.$$
$$\frac{\widetilde{p}_k}{p_k} \geq \frac{\mu(A)}{\widetilde{\mu}(\widetilde{A})} min_{j \in \Gamma} \frac{\widetilde{p}_j}{p_j} \text{ for } \mu(A) < \frac{\mu(A)}{\widetilde{\mu}(\widetilde{A})}.$$

Here $\mu(A), \widetilde{\mu}(\widetilde{A})$ are the principal eigenvalues of the matrices $A, \widetilde{A}$.

However, for $\mu(A) = \widetilde{\mu}(\widetilde{A})$ we cannot assert a definite result. More importantly, this proposition only states the price effects under the prescribed values of the principal eigenvalues without specifying the actual rate of profit. However, our price system also depends on the actual rate of profit.

[15]See Wilkinson (1965, pp. 88–89) and Strang (1980, pp. 284–285). As for a further analysis of perturbation, see e.g. Koyarik and Sherif (1985).

We return to the diagonalizable system and explore the relationship between the original matrix and the perturbed matrix by the following lemma[16]:

Proposition 5. *Let the matrix $A(\varepsilon) = A + \varepsilon F$ be diagonalizable for every complex number ε. Then all eigenvalues of $A(\varepsilon)$ are linear, that is of the form $\mu_i + \varepsilon a_i$. Here the eigenvalues μ_i, belong to the matrix.* (Kato 1982, p. 96; A Theorem of Motzkin-Taussky).

We should notice that is not restricted to be sufficiently small in this theorem. We now refer to the Sraffa–Leontief price system as discussed in the previous section. Interestingly, this proposition might be conformable to Goodwin's original conjecture on general coordinates referred to in Sect. 2.2, if eigenvalues of the linear form $\mu + \varepsilon a$ have had all positive a. In general, however, this if-clause cannot be valid. Applying Proposition 5 to the Sraffa–Leontief price system, we show an interesting theorem on technological changes. In this case, a perturbation matrix F may be interpreted as a pattern of technological changes, whose intensity is expressed by ε.

Theorem 5. *Suppose that the unperturbed input matrix A and the perturbed matrix $A + \varepsilon F$ have all simple roots distinct. Let the principal eigenvalues of the perturbed input matrix $A + \varepsilon F$ be $\mu(A) + \varepsilon a$. Then it follows:*

(i) *A price system of the perturbed Sraffa–Leontief system $[A + \varepsilon F, a^0]$ has a non-negative price if and only if*

$$1/(1+r) > \mu(A) + \varepsilon a \text{ for a given } r.$$

(ii) *The general solution of the perturbed Sraffa–Leontief price system will be of the form:*

$$p(r) = \sum_{i=1}^{M} e^{m_i(\varepsilon) r} c_i(\varepsilon) u_i, \tag{2.24}$$

where

$$M(\varepsilon) = [m(\varepsilon)_{i\ diag}] = \left[\left\{ \frac{1}{\mu_i + \varepsilon a_i} - (1+r) \right\}_{diag} \right]^{-1}.$$

[16]It is verified that our regular production system is structurally stable under perturbation. It is also shown that a system around the neighborhood of a given regular system is regular against the same primary factor. See Theorem 3 in Aruka (1990).

Aruka (1989) established for the Sraffa–Leontief price system a Nash Equilibrium for a given rate of profit r. By virtue of Theorem 5(i). we can see that for the regular system defined above there exists another Nash Equilibrium point for a new r' when r is changed slightly to r'. Thus our perturbed version of the Sraffa–Leontief price system could be extended to describe the relationship between choice of techniques and Nash solutions.

We suppose that the eigenvectors of a perturbed system only behave slightly: $u(\varepsilon)\approx u$. It may be also assumed that $H(\varepsilon)\approx H$. Hence, as for the constants determined by the initial conditions, $c(\varepsilon)\approx c$. If these approximations are permissible, we may conclude the following result on the price movement induced by technological change pattern F. The difference of the general solutions of the unperturbed system and the perturbed system, namely the price variation, may be estimated to be

$$\sum_{i=1}^{M} e^{(m_i(\varepsilon)-m_i)r} c_i u_i = \sum_{i=1}^{M} e^{\frac{\varepsilon a_i}{\{1-(\mu_i+\varepsilon a_i)(1+r)\}\{1-\mu_i(1+r)\}}} c_i u_i .$$

2.7 Conclusion

Let us turn to general coordinates again. The following relationship holds in Goodwin's general coordinates of the Sraffa–Leontief price system:

$$\left\{a_i^i x - x^t A x\right\},$$

We can have the inverse transformations $[I - (1+r)A]^{-1}$, H^{-1} under some reasonable assumptions. Once a configuration H has been adopted, the converse relationship also holds. In other words, the commutative relationship is always held. This situation is called "diffeomorphism" in mathematics. The use of "diffeomorphism" is often useful for analytical science. Goodwin's use of general coordinates are justified by diffeomorphism. There were some overstatements in Goodwin's original proposals. This, however, does not invalidate the usefulness of the use of general coordinates.

Diffeomorphism in the Sraffa–Leontief system requires a seemingly restrictive condition. In this article, some economic meanings of the condition have been established. We may be encouraged to use general coordinates in the light of these results.

Appendix

Proof of Proposition 3. Let μ_i be eigenvalues of a square matrix A of order n. Then we have a triangularization:

$$H^{-1}AH = [\mu_{i\ diag}].$$

We also define such a diagonal matrix G that its diagonal elements are $g_{ii}=\tau i (i=1, \ldots, n)$. Here we choose any τ fulfilling

$$0 < n\tau < \min_{\mu_i \neq \mu_j} |\mu_i - \mu_j|.$$

If $\mu_1 = \mu_2 = \ \ldots = \mu_{\mathrm{n}}$, we may choose τ arbitrarily. Furthermore, define $F = HGH^{-1}$. Thus it follows that

$$H^{-1}AH + \varepsilon G = H^{-1}[A + \varepsilon F]H = [\mu_i + \varepsilon g_{ii\ diag}]$$

for a sufficiently small $\varepsilon \neq 0$. The diagonal elements by definition are all distinct. Hence, any matrix can be made to have all distinct eigenvalues through fine tuning □

Proof of Theorem 1. Since the components of $R(g, g')$, a_i, b_i are determined by ε and f_{kl}. $R(g, g')$ can be expressed a polynomial of ε and f_{kl}. Arranging this polynomial with respect to ε for some $m \geq n^2 - 1$,

$$\upsilon_0(f_{kl})\varepsilon^m + \upsilon_1(f_{kl})\varepsilon^{m-1} + \cdots + \upsilon_{m-1}(f_{kl})\varepsilon + \upsilon_m(f_{kl}) = 0.$$

Since ε is a sufficiently small non-zero number, $R(g, g') = 0$ holds only if

$$\upsilon_0(f_{kl}) = 0, \upsilon_1(f_{kl}) = 0, \ldots, \upsilon_{m-1}(f_{kl}) = 0, \upsilon_m(f_{kl}) = 0.$$

These are the condition that $A + \varepsilon F$ has multiple roots for a sufficiently small non-zero number ε. These are not identically fulfilled with any perturbation direction F, which, as a whole, constitutes a projective space of dimension $n^2 - 1$. Only a perturbation direction F specified by the above condition gives multiple roots. Thus such a perturbation direction determines an algebraic manifold V of a smaller dimension than $n^2 - 1$ in a whole projective space. Excepting these singularities, eigenvalues of $A + \varepsilon F$ are all distinct for any perturbation direction F. A graph of V is of measure zero. Hence we have all distinct eigenvalues almost everywhere. □

Proof of Corollary 1 to Theorem 1. $\upsilon_m(f_{kl})$ is really det|A|. If A has no repeated roots, det|A| does not vanish. Hence $\upsilon_m(f_{kl})$ does not vanish. $A + \varepsilon F$ then will not have repeated roots for a sufficiently small $\varepsilon(\neq 0)$. □

Proof of Theorem 3. In matrix the differential equation system is of the form:

$$\frac{dp}{dr} = Ap + (1+r)A\frac{dp}{dr}. \tag{2.25}$$

This is reformulated into the following form:

$$\begin{aligned}\frac{dp}{dr} &= \{[I-(1+r)A]^{-1}A\}p \\ &= \{[I-(1+r)A]^{-1}(A^{-1})^{-1}\}p. \\ &= \{A^{-1}[I-(1+r)A]\}^{-1}p \\ &= [A^{-1}-(1+r)I]^{-1}p.\end{aligned}$$

By the assumptions, $[I-(1+r)A]$ is non-negatively invertible because of the Frobenius Theorem. Then it is expanded in the following series:

$$[I-(1+r)A]^{-1} = I + (1+r)A + (1+r)^2A^2 + \ldots \tag{2.26}$$

Employing the expansion, it is verified that $[I-(1+r)A]^{-1}A$ and A are commutable:

$$[I-(1+r)A]^{-1}A \cdot A = A \cdot [I-(1+r)A]^{-1}A \text{ for } r \leq R. \tag{2.27}$$

Therefore, by virtue of Lemma matrix $[I-(1+r)A]^{-1}A$ has the same eigenvectors with a commutable matrix A. We denote $[I-(1+r)A]^{-1}A$ by B. Let matrix H be constructed by the eigenvectors u of matrix A. The general solution of this differential equation system can then be of the form[17]:

$$p(r) = e^{Br}p(0).$$

If A is non-singular, it follows that

$$H^{-1}AH = [\mu_{i\ diag}],$$

from which it is easily transformed:

$$A^{-1} = H^{-1}[\mu_{i\,diag}]^{-1}H = H^{-1}\left[\frac{1}{\mu_i}{}_{diag}\right]H. \tag{2.28}$$

---

[17] $p(0)$ is a solution of (2.12) when r is set to zero:
$p(0) = [I–A]\ a^0$.
This is nothing but a vector of labor values.

By making use of (2.15) and (2.28), the similarity transformation of $[I-(1+r)A]^{-1}A$ is derived in the following manner:

$$\begin{aligned}[I-(1+r)A]^{-1}A &= [A^{-1}-(1+r)I]^{-1}\\ &= \left[H\left[\frac{1}{\mu_i}{}_{diag}\right]H^{-1}-(1+r)I\right]^{-1}\\ &= H\left[\left[\frac{1}{\mu_i}{}_{diag}\right]-(1+r)I\right]^{-1}H^{-1}.\end{aligned}$$

Accordingly,

$$H^{-1}[I-(1+r)A]^{-1}AH = \left[\frac{1}{\mu_i}-(1+r)_{diag}\right]^{-1} = M. \qquad (2.29)$$

The general solution then is represented as

$$p(r) = e^{Br}p(0) = He^{Mr}H^{-1}p(0) = \sum_{i=1}^{M} e^{m_i r}c_i u_i,$$

where $M = [m_{i\,diag}] = \left[\frac{1}{\mu_i}-(1+r)_{diag}\right]^{-1}$, and c_i, is a constant determined by the initial condition $c = H^{-1}\,p(0)$.

Applying $p = e^{mr}u$ to (2.25) as a particular solution, it follows that

$$mp = [A^{-1}-(1+r)I]^{-1}p.$$

This is arranged as follows:

$$Ap = \frac{m}{1+m(1+r)}p.$$

By replacing $\frac{m}{1+m(1+r)}$ with μ,

$$Ap = \mu p.$$

Thus it has been proved that the eigenvalues m of the differential system correspond to the eigenvalues μ of the Sraffa–Leontief price system by the relationship of $\frac{\mu}{1-\mu(1+r)}$. □

Proof of Corollary 2. In the assumptions of Theorem 3. the principal eigenvalue is only one real value. Choose r closely to the maximum rate of profit R defined by the principal eigenvalue $\mu(A) = 1 + R$. Then, in the principal eigensector, the corresponding component of M in (2.29) will be:

$$\left\{\frac{1}{\mu(A)} - (1+r)\right\}^{-1} = \frac{\mu(A)}{1 - \mu(A)(1+r)} > 0,$$

since $\frac{1}{1+r} > \mu(A) > 0$. The particular solution of the principal eigensector will definitely be divergent. In a productive system, all the eigenvalues are smaller than 1 in terms of absolute values:

$$1 > |\mu_i|.$$

Consequently, the general solution will be unstable as the rate of profit becomes the maximum rate of profit R. □

Proof of Corollary 3. In our price movement equation system the initial value is the value when r is set to 0:

$$p(0) = [I - A]^{-1} a^0,$$

which can be expanded in the following manner, given the assumptions of Theorem 3:

$$p(0) = a^0 + Aa^0 + A^2 a^0 + \dots .$$

Taking account of our requirement $Aa^0 = \mu a^0$, this will furthermore be rearranged into

$$p(0) = (1 + \mu + \mu^2 + \dots) a^0 = \frac{1}{1-\mu} a^0.$$

Then it follows that

$$p(r) = \frac{1}{1-\mu} H e^{Mr} H^{-1} a^0.$$

H is constructed by the columns of eigenvectors u of A. Substitute a^0 into u_1. The initial value will be expressed as follows:

$$c = H^{-1} a^0 = [1, 0, \dots, 0]',$$

because of $H^{-1}H = I$. Hence other coefficients other than the first component of c are 0.

Therefore

$$p(r) = \sum e^{m_i r} c_i u_i = e^{m_1} u_1 = e^{m_1} a^0.$$

This completes the proof. □

Proof of Theorem 4. Let m be the ratio satisfying the condition

$$m = \frac{a_i p}{a_i^0} = \frac{a_i A p}{a_i a^0} = \frac{a_i A^2 p}{a_i A a^0} = \ldots \text{ for } r \in (0, R].$$

This ratio implies "the balancing proportion" for the prices against the changes of the rate of profit. The activity vector s preserving this critical proportion has to satisfy the following condition:

$$m = \frac{s A^n p}{a_i A^{n-1} a^0} \text{ for } r \in (0, R]. \tag{2.30}$$

Substituting (2.12) into (2.30), it follows that, taking account of (2.27),

$$s\{[I - (1+r)A]^{-1} A - mI\} A^{n-1} a^0 = 0.$$

Let z be the vector $s\{[I - (1{+}r)A]^{-1} A - mI\}$ and L the matrix $[a^0, Aa^0, A^2 a^0, \ldots, A^{n-1} a^0]$:

$$zL = 0. \tag{2.31}$$

Hence if L is of dimension n, the null space of L is of dimension $n{-}n = 0$. Thus (2.31) has only the solution $z = 0$. Conversely, if $z = 0$, the dimension of L is n. L of dimension n implies that

$$a^0, Aa^0, A^2 a^0, \ldots, A^{n-1} a^0$$

are linearly independent. $z = 0$ means that s is a left-hand eigenvector of matrix $[I - (1{+}r)A]^{-1} A$:

$$s[I - (1+r)A]^{-1} A = ms,$$

i.e.

$$s[A^{-1} - (1+r)A] = ms.$$

This completes the desired results:
$sA = s\frac{m}{1+m(1+r)}$ for a left-hand eigenvector s,

$Au = \frac{m}{1+m(1+r)}$ for a right-hand eigenvector u.

The second result is just the same as Theorem 3(ii). Thus the necessary and sufficient condition of this theorem is equivalent to normal modes of the Sraffa–Leontief differential equation system (2.14).

Finally, it remains to examine the assumptions $sa^0 \neq 0$, and $a^0 \neq (1+R)Aa^0$. It is easily verified that if $sa^0 \neq 0$, $a^0 \neq (1+R)Aa^0$, then $a^0, Aa^0, A^2a^0, ..., A^{n-1}a^0$ are linearly independent. Suppose, on the contrary, they are linearly dependent. In the case $n = 2$, appropriating the non-zero numbers 1 and $-(1+R)$, it follows that $a^0-(1+R)Aa^0=0$. a^0 happens to be a left-hand eigenvector of A. Then, by orthogonality of the left- and right-hand eigenvectors in an eigenvector space, $sa^0 = 0$. Conversely, if $sa^0 = 0$, a^0 is an eigenvector of A. While $z \neq 0$. In the case $n = 2$, then, in view of Jordan normal forms, it can be expressed:

$$s^1 = ms^1 + s^2.$$

Here s^i is a generalized eigenvector. Post-multiplying the left-hand vector a^0 on both sides, it follows that

$$s^1a^0 = ms^1a^0 + s^2a^0.$$

This implies s^2a^0 is 0 if s^1a^0 is 0. By induction on a cyclic subspace, it is proved that if a^0, $a^0, Aa^0, A^2a^0, ..., A^{n-1}a^0$are linearly independent, $sa^0 \neq 0$ (rsp. $a^0 \neq (1+R)Aa^0$). □

Proof of Proposition 4. By one of the Frobenius theorems, the principal eigenvalues $\mu(A), \widetilde{\mu}(\widetilde{A})$ fulfilling $Ap = \mu(A)p, \widetilde{A}\widetilde{p} = \widetilde{\mu}(\widetilde{A})\widetilde{p}$ are positive and simple. Suppose, on the contrary, that for $\mu(A) > \widetilde{\mu}(\widetilde{A})$,

$$\frac{\widetilde{p}_k}{p_k} > \frac{\mu(A)}{\widetilde{\mu}(\widetilde{A})} max_{j\in\Gamma} \frac{\widetilde{p}_j}{p_j}.$$

Then it follows that

$$\frac{\mu(A)}{\widetilde{\mu}(\widetilde{A})} > max_{j\in\Gamma} \frac{\widetilde{p}_j/p_j}{\widetilde{p}_k/p_k} > 1.$$

Hence $\mu(A) < \widetilde{\mu}(\widetilde{A})$. This is a contradiction. Similar to the above, we can prove:

$$\frac{\widetilde{p}_k}{p_k} \geq \frac{\mu(A)}{\widetilde{\mu}(\widetilde{A})} min_{j\in\Gamma} \frac{\widetilde{p}_j}{p_j} \text{ for } \mu(A) < \widetilde{\mu}(\widetilde{A}).$$

□

Proof of Theorem 5. (i) If it is the case that

$$\mu(A) + \varepsilon a > \frac{1}{1+r} > \mu(A),$$

a price system will be made unreasonable for a given rate of profit r. In other words, the reasonableness of a price system after perturbation will depend on not only a matrix F but also on the magnitude of a given rate of profit. Thus Proposition 5 supports statement (i).[18]

(ii) The perturbed matrix $A + \varepsilon F$ is assumed to be non-singular. Define $H(\varepsilon)$ to be a transformation matrix being constructed by the eigenvectors $u(\varepsilon)$ of $A + \varepsilon F$. All the arguments in Theorem 3 will be valid, if A is replaced with $A + \varepsilon F$ as well as μ with $\mu + \varepsilon a$. Hence

$$\begin{aligned} p(r) &= H(\varepsilon)e^{M(\varepsilon)r}H(\varepsilon)^{-1}p(0) \\ &= \sum_{i=1}^{M} e^{m_i(\varepsilon)r} c_i(\varepsilon)u_i, \end{aligned}$$

where

$$M(\varepsilon) = [m(\varepsilon)_{i\,diag}] = \left[\left\{\frac{1}{\mu(\varepsilon)_i} - (1+r)\right\}_{diag}\right]^{-1},$$

$c(\varepsilon)$ is a constant determined by the initial condition $c(\varepsilon) = H(\varepsilon)^{-1}p(0)$. □

References

Aruka Y (1987) Generalized Goodwin's theorem in the linear multisectoral model. Research Papers No. 9. The Institute of Business Research, Chuo University (Presented at the Far Eastern Meeting of the Econometric Society, Tokyo, Japan)

Aruka Y (1988) General coordinates and price dynamics in the multisectoral model. Research Papers No. 11. The Institute of Business Research, Chuo University (Presented at the Australasian Meeting of the Econometric Society, Canberra, Australia)

Aruka Y (1989) Perturbed version of the Leontief price system and the Nash equilibrium. Research Papers No. 13. The Institute of Business Research, Chuo University

[18] A study of this issue had been done in the context of a Nash equilibrium elsewhere. See Aruka (1989).

Aruka Y (1990) Perturbation theorems on the linear production model and some properties of eigenprices. Working Paper in Economics and Econometrics, Australian National University, No. 203
Benavie A (1972) Mathematical techniques for economic analysis. Prcntice-Hall, Englewood Cliffs, NJ
Blatt JM (1983) Dynamic economic systems. M.E. Sharpe, New York
Dietzenbacher E (1988) Perturbations of matrices: a theorem on the Perron vector and its applications to input-output models. J Econ 48:389–412
Gantmacher FR (1959) The theory of matrices, vol I. Chelsea, New York
Goodwin RM (1953) Static and dynamic linear general equilibrium models. In: Netherlands Economic Institute (ed) Input–output relations. Netherlands Economic Institute, The Netherlands (also reprinted in Goodwin 1983a)
Goodwin RM (1976) Use of normalized general coordinates in linear value and distribution theory. In: Polenske K and Skolka J (eds) Advances in input–output analysis (also reprinted in Goodwin 1983a)
Goodwin RM (1983a) Essays in linear economic structures. Macmillan, London
Goodwin RM (1983b) Capital theory in orthogonalised general coordinates. In: Essays in linear economic structures. Macmillan, London
Goodwin RM (1983c) Disaggregating models of fluctuating growth. In: Goodwin RM, Krueger M, Vercelli A (eds) Nonlinear models of fluctuating growth. Springer, Berlin
Goodwin RM (1986) Swinging along the Turnpike with von Neumann and Sraffa. Camb J Econ 10:203–210
Goodwin RM, Punzo LF (1987) The dynamics of a capitalist society. Polity, Basil Blackwell, Cambridge
Kan T, Iri M (1982) Jordan Hyojunkei (Jordan Normal Forms). University of Tokyo Press, Tokyo (in Japanese)
Kato T (1982) A short introduction to perturbation theory for linear operators. Springer, NewYork
Klein E (1973) Mathematical methods in theoretical economics. Academic, New York
Koyarik ZV, Sherif N (1985) Perturbation of invariant subspaces. Linear Algebra Appl 64:93–113
Mas-Colell A (1985) The theory of general economic equilibrium: a differential approach. Cambridge University Press, New York
McKenzie L (1960) Matrices with dominant diagonals and economic theory. In: Arrow KJ, Karlln S, Suppes P (eds) Mathematical methods in the social sciences 1959. Stanford University Press, Stanford
Miyao T (1977) Generalization of Sraffa's standard commodity and its complete characterization. Int Econ Rev 18:151–162
Nikaido H (1968) Convex structures and economic theory. Academic, New York
Satake I (1974) Senkei Daisugaku (Linear algebra). Shokabo, Tokyo (in Japanese)
Schefold B (1976) Relative prices as a function of the rate of profit. Zeitschrift fuer Nationaloekonomie 36:21–48
Schefold B (1978) On counting equations. Zeitschrift fuer Nationaloekonomie 38:253–285
Seneta E (1981) Non-negative matrices and Markov chains, 2nd edn. Springer, New York
Sraffa P (1960) Production of commodities by means of commodities. Cambridge University Press, Cambridge
Strang G (1980) Linear algebra and its applications, 2nd edn. Academic, Orlando
Wilkinson JH (1965) The algebraic eigenvalue problem. Clarendon, Oxford

Chapter 3
Possibility Theorems on Reswitching of Techniques and the Related Issues of Price Variations

Yuji Aruka

3.1 Some Difficulties on the Factor–Price Frontier and the Wage Curve

Aftermath to a great debate called *capital controversy* over three decades ago, several seemingly useful results on production theory so far became familiar with us as given by Diewert (1982), Newman (1987), and others: GNP functions, Hotelling's lemma and Shepard's lemma. Intermediate microeconomic textbooks usually collect the topics, although these results would virtually be invalid in front of a combination of the use of intermediate goods like in the Sraffa–Leontief system and an introduction of joint-outputs, even if the neoclassical distribution principle were employed. Exception for the case is only a kind of *the factor price equalisation theorems*, as Woodland (1982) showed.[1] Many economists, however, do not feel serious about those limitations of neoclassical production theory. We may wonder whether they have given up working with those subjects or discarded a theory of production.

On the other hand, the Sraffian model may also encounter a serious difficulty to describe reswitching on the wage curve, i.e., the factor–price frontier, because our measurement of real wage in terms of commodities weighted by prices is to be accompanied with price effects, whose penetration may intervene erratically a relation between the rate of profit and the rate of real wage. We can in fact observe a perverse case in which there is a reswitching pair of technique between two

[1]See Chap. 5 titled "Intermediate Inputs and Joint Outputs" in Woodland (1982). Specifically, see p. 140.

Reprinted from Bulletin, Institute of Economic Research, Chuo University 30, Aruka, Y., Possibility Theorems on Reswitching of Techniques and the Related Issues of Price Variations, 79–119 (2000). With kind permission from Chuo University, Tokyo.

Y. Aruka
Faculty of Commerce, Chuo University, Higashinakano, Hachioji, Tokyo 192-0393, Japan
e-mail: aruka@tamacc.chuo-u.ac.jp

Y. Aruka (ed.), *Complexities of Production and Interacting Human Behaviour*,
DOI 10.1007/978-3-7908-2618-0_3, © Springer-Verlag Berlin Heidelberg 2011

alternative production systems but does not appear any double crossing of both distinct wage curves under comparison.[2]

Index problems on the side of production deserve of being studied to solve such complications. The development of theory has been kept continued by Fisher and Shell (1972, 1998). The Sraffian theory may be usefully extended to comprehend their theory. At this stage when several shortcomings were identified for both theories, much superficially distance between neoclassical economics and Sraffian economics would be reduced, despite of their fundamental difference. Thus an instruction will be prescribed to contribute a new development for production theory.

I can state that the fundamental difference between both theories is no less than the way how to incorporate *surplus* into an economic system. The other differences such as the distribution principles as well as the use of intermediate goods may be rather considered minor by the following arguments on appearance of reswitching: It sometimes seemed us important whether the neocalssical distribution principle or the Sraffian distribution principle is to be employed.[3] It may however be superficial, because many writers on reswtiching often devoted themselves to working with the neoclassical distribution principle in order to detect reswitching. Exception was an Austrian roundabout process without any sectoral interrelation as employed by Samuelson (1966). Given any sectoral interrelation of production within a system, it is rather deemed ironical to prove that appearance of reswitching is less probable in the Sraffian distribution principle than in the neoclassical distribution principle. It immediately turns out if the two-sector model is employed for their comparison.

Although the present paper is only concerned with reswitching mechanism by restricting the analytical domain onto the two-sector production model, the proof of probability on reswitching is one of our main results in this paper: The neoclassical distribution principle has much more influence on appearance of reswitching than the Sraffian distribution principle, whether the employed model is a model omitting the use of intermediate goods or not. This clarifies why we have *a fortiori* preferred the Sraffian joint-production model to illustrate reswtiching. The reason may be unambiguous if a joint-production model has appropriately been transformed into a vertically integrated system as Pasinetti (1977, 1981) recommended, because the latter system of a joint-production system formally coincides with an imaginary single production system under the neoclassical distribution principle. Put another way, at first, a reswitching link is *via* a choice of distribution principle. More

[2]See Figs. 3.7 and 3.8 in Sect. 3.2.

[3]Imagine the production function with a property of constant return on scale and apply the Euler rule to it. Then it follows an exhaustive distribution of net outputs by marginal productivity on each factor of production. We call this *the neoclassical distribution principle* since Wicksell, or *the exhaustive distribution principle*. Even if there came out residuals without constant return, a rental price as marginal productivity would not be given up in the neoclassical arguments. Thus it does not matter to collect depreciation on capital when discussing the factor-price frontier. Attention was paid only on the item rAp, for instance. On the contrary, recovery for capital inputs all is presumed *in advance* in the Sraffian model:$(1+r)Ap$. The difference may be represented by the difference of price-cost equations:

$rAp+wl=p$ in a neoclassical model while $(1+r)Ap+wl=p$ in the Sraffian model,

as described in details in a later section of the present paper Sect. 3.2.

importantly, it has been verified that the neoclassical principle is *a fortiori* committed to generate reswitching. Thus the distributive link may be a decisive hindrance for the neoclassical factor–price frontier, as we anticipated.

There would however remain to exist a serious gulf on understanding the factor–price frontier between the standard economic theory and the Sraffian theory, even if we dealt with the two-sector model.

Summarising these, what it seems us to be essentially fundamental in the filed of production theory is neither to do with the different distribution principle nor with the use of intermediate goods. It essentially rather matters to do with *surplus of production*. Rate of interest or profit in the neoclassical theory is irrelevant to surplus outputs in real terms of an economy. The rate of profit appearing in the factor–price frontier of the Sraffian theory is not the same as a price of capital as a factor of production. The former rate reflects a distributive claim on how to divide surplus, which is socially determined. Surplus as a whole is technologically determined. In other words, the Sraffian theory does not have a direct link from a distributive factor such as rate of profit to a certain technological implication. We must be careful for the difference of implications on the factor–price frontier /or the wage curve although nomination is the same. Employing activity analysis, this situation can be concisely illustrated by virtue of Figs. 3.1 and 3.2 as shown in Scehfold (1989).[4] Our rate of profit definitely is different from a rent price of capital in view of Fig. 3.2.[5]

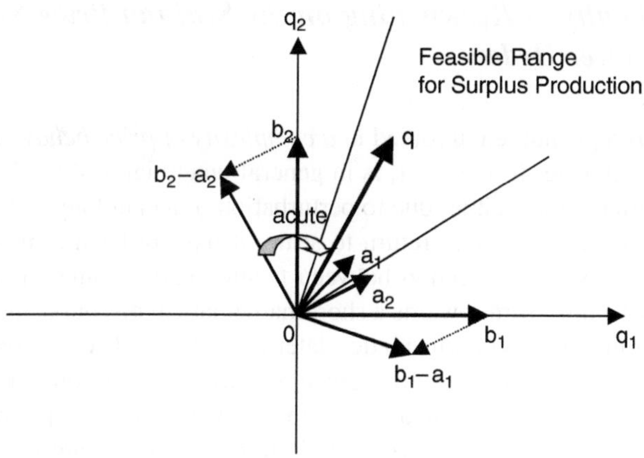

Fig. 3.1 An activity analysis for surplus production

[4]See Part II-10 "Graphic Techniques Method" in Schefold (1989) where many variations of joint productions exhaustively are mentioned. a_i denotes an input vector, b_i an output vector. Positive surplus for all goods requires the condition of inner product $b_i - a_i \cdot q > 0$ for all i which means the angle between $b_i - a_i$ and q is *acute*. In Fig. 3.2, it is noted that it usually only matters the hyperplane of (labour), i.e., $l_1 + l_2 = 1$, while, in the Sraffian model, the hyperplanes of $r = 0$ and $r = R$ (maximum) also equally exert the decisive effect on distribution of an economy.

[5]Incidentally, it is trivially evident that future production decisively depends on the way of distribution of real surplus in any historical era of production. An economy without surplus but

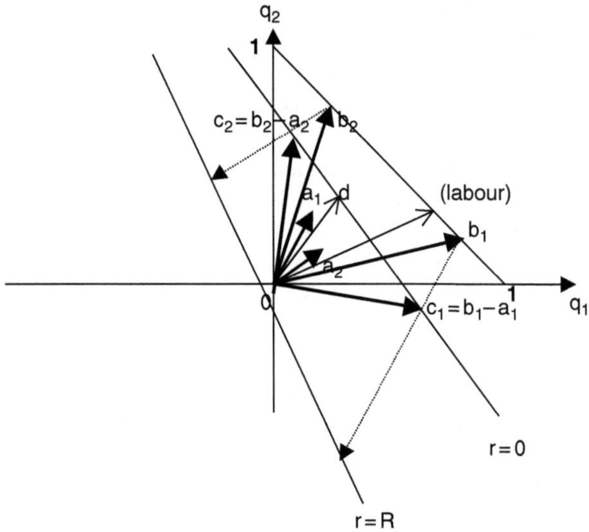

Fig. 3.2 Distributive relations given demands and labours in the activity analysis

3.1.1 Difficulty of Reswitching on the Sraffian Price System of Lower Order

Reswitching is a phenomenon rooted in *a complexity of price behaviour* in context of distributive change. It, however, is in general unambiguously indistinguishable to know the price effects either due to perturbation of technology or to a change of rate of profit, given a constant return to scale. It has not been a main concern to analyse the effects of intersection between technological change and distributive change. Thus we must turn towards a short mathematical discourse in the new light of linear mapping to do with this matter, later in Sect. 3.3. The intersection of both will be analysed with resort to convergence property of hyperbolic linear mapping.

The association of net-producibility with the diagonal assumptions guarantees a tractable treatment of price movements in terms of rate of profit r, as shown in Aruka (1991). Leaving the assumption of net-producibility aside, the Sraffian price system by itself is confronted with a discontinuity problem. If a rate of profit r to

growing would be an awkward one even if it were a fictitious artifice. Classical economists gave much consideration on this aspect to have their unique views of capitalistic evolution, without being confining themselves into strictly limited issues. How insightful it was has invoked our concern on classical economics, which has at present been inherited by a new stream called *Evolutionary Economics*. This view is giving us a comprehensive view of capitalistic dynamics. In this stream, a series of neoclassical propositions can be seriously tested against facts. The works have just been undertaken. See Introduction of Nelson (1996), for instance.

prices p relationship were of hyperbolic form on the r–p plane in the range of smaller part of r, a similar system could exceed the asymptotic vertical line of infinity on the r–p plan once a possible cross with the former price curves $p(r)$ has done. We can easily illustrate this by means of a numerical example of 3 processes by 3 goods economy. This kind of discontinuity can occur even if the absolute eigenvalues are all less than 1.

Let an input matrix A, a no joint-output I be as follows:

$$A = \begin{bmatrix} 0.4 & 0.1 & 0.2 \\ 0.2 & 0.1 & 0.1 \\ 0.09 & 0.6 & 0.38 \end{bmatrix}, \quad I = \begin{bmatrix} 1 & 0 & 0 \\ 0 & 1 & 0 \\ 0 & 0 & 1 \end{bmatrix}.$$

Since eigenvalue λ of matrix A is

$$\lambda = \{0.706612, 0.0866941 + 0.0443396i, 0.0866941 - 0.0443396i\},$$

their absolute values are *all less than* 1:

$$|\lambda| = \{0.706612,\ 0.0973749, 0.0973749\}.$$

Let a labour vector be:

$$l = (0.3, 0.4, 0.3)'$$

By the use of both matrices A and I, we set out matrix C containing parameter r:

$$C(r) = I - (1 + r)A$$

Then C can be written as follows:

$$C = \begin{bmatrix} 1 - 0.4\,(1 + r) & -0.1\,(1 + r) & -0.2\,(1 + r) \\ -0.2\,(1 + r) & 1 - 0.1\,(1 + r) & -0.1\,(1 + r) \\ -0.09\,(1 + r) & -0.6\,(1 + r) & 1 - 0.38\,(1 + r) \end{bmatrix}$$

The price is solved as

$$p = C^{-1}l.$$

The solution is of the form of rational polynomial (Fig. 3.3):

$$p_1 = 8(-29r^2 - 3r - 349)/(67r^3 - 1119r^2 + 6361r - 2453)$$

$$p_2 = 5(-67r^2 + 310r - 423)/(67r^3 - 1119r^2 + 6361r - 2453)$$

$$p_3 = 3(117r^2 - 36r - 1213)/(67r^3 - 1119r^2 + 6361r - 2453)$$

Fig. 3.3 Price functions of rational polynomial form

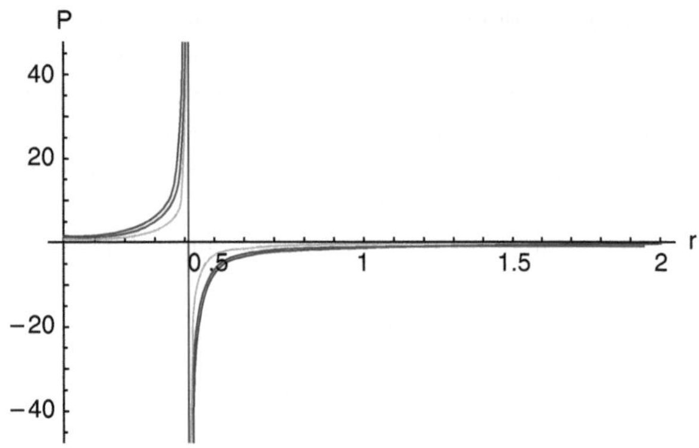

This is an example that a strong hyperbolicity sometimes generates discontinuity before an opportunity of a reswitching point is coming out. It can however be proved that the discontinuity as shown in the figure attributes to a special form of hyperbolicity often taken place in a three-dimensional case. One reason for this happening is that a relative price in the 3 by 3 case will be reduced to essentially a rational quadratic form. If we move our domain at least to 4 processes by 4 goods economy, a form of relative price will be of more complex. Hyperbolicity in the essentially over three-dimensional case will then contribute an existence proof for reswitching. Thus we can anticipate that there may be some resistance against reswitching happening as long as the Sraffian price system of *lower order* is considered.

3.2 Possibility Theorems on Reswitching in the Two-Sector Models

We now argue our issues of reswitching only in the domain of two-sector model with the use of intermediate goods like in the Sraffa system. There are some essential variants even in the two-sector model. It is often used to adopt a certain custom to confirm an occurrence of reswitchng because of an easy joy to show it. Put it another way, we have often argued it in the neoclassical tradition that the factor income of the economy must be completely distributed without residuals. In short, we have utilised the hypothesis of "exhaustive distribution among factors of production". The simplest way to find an example in the following two-sector system of the capital good industry denoted by index 1 and the consumer good industry by index 2:

$$\begin{aligned} rp_1k_1 + w &= p_1y_1, \\ rp_1k_2 + w &= p_2y_2. \end{aligned} \tag{N}$$

These show the factors cost equation for the capital good as well as for the consumer good producing industry. Here k_i is the capital–labour ratio, y_i the output–labour ratio, p_i the price in the ith industry. Now pk_i must be regarded as the value of capital. Also r is to be identified with the rate of profit, w with the rate of wage. So rpk_i implies the capital income per capita labour while w the labour income per capita labour. A production function with constant returns to scale is behind our distributive setting here.

$$y_i/l_i = f(a_{i1}/l_i, l_i) \text{ for } i = 1,2,$$

where a_{i1} is the capital input, l_i the labour input per capita output for the i-th industry. It holds, indeed:

$$k_1/y_1 = a_{11}/l_1,\ k_2/y_2 = a_{21}/l_2$$

Thus we may reformulate this picture in terms of the previous notation of input coefficients a_{ij}:

The unit activity of production for capital good: $\{a_1 = (a_{11}, a_{21}) = (a_{11}, 0), l_1\} \to (1, 0)$

The unit activity of production for consumer good: $\{a_2 = (a_{21}, a_{22}) = (a_{21}, 0), l_2\} \to (0, 1)$

In addition, we introduce a normalisation:

$$l = (l_1, l_2)' = (1, 1)'$$

We therefore easily check the result:

$$k_1 < k_2, \Leftrightarrow a_{11} < a_{21}\, ; k_1 > k_2, \Leftrightarrow a_{11} > a_{21}$$

Then the above (**N**) system is equivalent to:

$$rp_1 a_{11} + w = p_1,$$
$$rp_1 a_{21} + w = p_2.$$

In view of these equation, the wage curve is to be derived as:

$$w(r) = -\frac{p_2(1 - a_{11}r)}{-1 + a_{11}r - a_{21}r}.$$

Hence it holds:

$$w'(r) > 0 \text{ for any nonnegative } r,$$

$$w''(r) > 0, \text{ if } a_{11} < a_{21}.$$

In other words, the wage curve of the (**N**) system will be always convex shaped towards the origin of the r–w plane when the capital good industry is more labour-intensive in her production than the consumer good industry. We call this curve α-curve. On the contrary, in the case of

$$a_{11} > a_{21},$$

we may encounter a nonconvex shaped curve. We call the latter one β-curve. It can be easily verified to check of the expression of $w''(r)$ whose denominator is negative around the origin, then the sign of the denominator may turn itself into the opposite one as r increases.[6] We can reproduce these arguments immediately in numerical terms. Suppose, for example, that a α-curve is composed by $a_{11} = 0.4$, $a_{21} = 0.6$, $p_2 = 1$, while a β-curve by $a_{11} = 0.5, a_{21} = 0.3, p_2 = 0.95$. It is apparent to see the two distinct intersections between the wage curves α, β (Fig. 3.4).

The above argument has been done with a specification that any consumer good a_{i2} must not be used as an intermediate good. An extension of our observation in this context could however be valid even if this kind of restriction were removed, unless we left from the exhaustive distribution principle. Reswitching phenomena will not depend on whether a production system under observation contains a basic good or nonbasics in the sense of Sraffa., as Levahri and Samuelson (1966) could not help consenting. Thus we examine a more general case of production system:

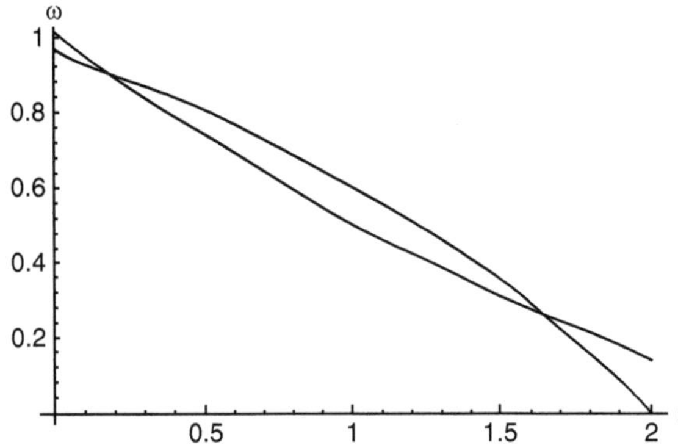

Fig. 3.4 A reswitching phenomenon in a two sector model

[6]The derivative of first order gives

$$w'(r) = \frac{a_{21}\,p_2}{(-1 + a_{11}\,r - a_{21}\,r)^2}.$$

The derivative of second order gives

$$w''(r) = \frac{2(a_{11} - a_{21})a_{21}p_2}{\{-1 + (a_{11} - a_{21})r\}^3}.$$

Hence, if $a_{11} < a_{21}$, it holds $w''(\,r\,) > 0$.

The first unit activity of production: $\{a_1 = (a_{11}, a_{21}), l_1\} \rightarrow (1, 0)$
The second unit activity of production: $\{a_2 = (a_{21}, a_{22}), l_2\} \rightarrow (0, 1)$
The system fulfils an indecomposable property of production relations among inputs used. It is noted that $A = (a_1, a_2)'$ is input matrix, $I = ((1, 0), (0, 1))'$ output matrix without joint-production. As the price equation under the neoclassical distributive principle is represented in matrix form:

$$rpA + wl = p,$$

the price solution is written in

$$p = [I - rA]^{-1} l.$$

This solution will be assured nonnegativeness around at $r = 1$ if $I - A$ is nonnegatively invertible.

3.2.1 *A Measurement Problem*

If we should cling to the principle of exhaustive distribution of factors income similarly as in the (**N**) system, we would have not found much burden to show reswtiching as a numerical exercise. Let two input matrices associated with two different system α, β be A^α, A^β, two labour input vector be l^α, l^β. We then assign them numerical values in the following way:

$$A^\alpha = \begin{bmatrix} 0.1 & 0.6 \\ 0.4 & 0.3 \end{bmatrix} \quad l^\alpha = \begin{bmatrix} 0.99 \\ 1 \end{bmatrix}; \quad A^\beta = \begin{bmatrix} 0.55 & 0.1 \\ 0.2 & 0.7 \end{bmatrix} \quad l^\beta = \begin{bmatrix} 1.01 \\ 1 \end{bmatrix}$$

We can check two distinct switch points at the corresponding positive rates of profit:

$$r_1 = 0.345686, \ r_2 = 0.847824$$

at which rates the prices are equally appreciated[7] (Fig. 3.5):

$$p^\alpha{}_1 = p^\beta{}_1.$$

It will be required a certain sophisticated artifice to give a wage curve in a general production system in which a number of intermediate goods simultaneously

7

$$p_1^\alpha = \frac{-4.71429 - 1.44286\,r}{-4.7619 + 1.90476\,r + r^2};\ p_1^\beta = \frac{2.76712 - 1.66301\,r}{2.73973 - 3.42466\,r + r^2}.$$

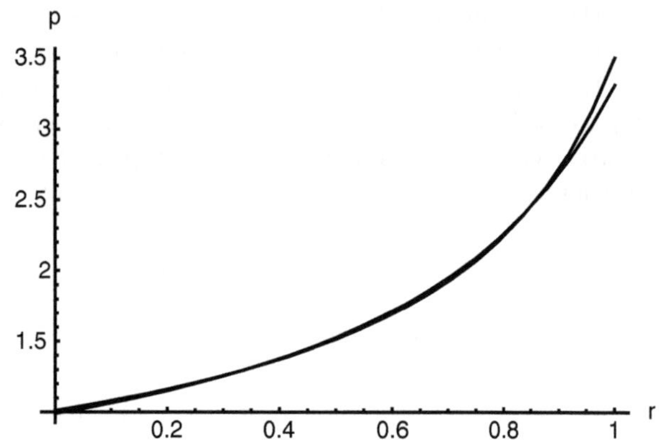

Fig. 3.5 A reswitching in terms of price function

be appropriated to consumer goods. A wage curve of the Sraffian price system may usually be defined as the formula of:

$$\omega(r) = 1/dp$$

Here

$$d = (d_1, d_2)$$

is considered a wage goods basket in which all entries of goods in the economy are permissible to enter. It is evident that this formulation can cause a so-called index or measure problem, even though a basket always is physically given fixed after prices change in our setting. The shape of wage curve may have a different form according to a different set of values of wage basket *d*. It may be possible that another selection of *d* can eliminate otherwise an occurrence of reswitching. We in fact have a disguised example of nothing of reswitching in the face of existence of it among the first processes of production. We employ the same numerical example in the above except for the value assignments of *d*. If *d* were set

$$d = (0.84,\ 0.16),$$

there would seemingly be no intersection of the wage curves under observation (Fig. 3.6).[8]

8

$$w^{\alpha} = \frac{1 - 0.4\,r - 0.21\,r^2}{0.9916 + 0.30188r};\ w^{\beta} = \frac{1 - 1.25r + 0.365\,r^2}{1.0084 - 0.5655\,r}.$$

The points fulfilling $w^{\alpha} = w^{\beta}$ contain the complex conjugates:

$$r_1 = -4.40937, r_2 = 0.463945 - 0.478274i, r_3 = 0.47854 + 0.478274i.$$

There however is not found any intersection on the r-ω plane (Fig. 3.6).

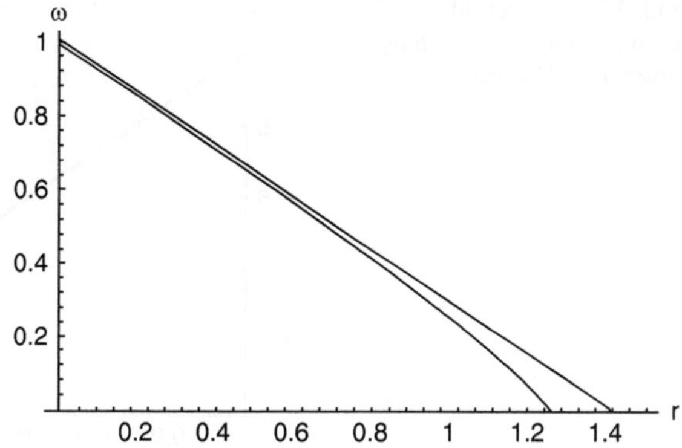

Fig. 3.6 A disguised example of non-reswitching given $d = (0.84, 0.16)$

If d were

$$d = (0.85, 0.15),$$

there could graphically be observed two intersection.[9] Calculation however indicates these points of intersection all are the complex conjugates (Fig. 3.7).

We could, of course, discover a normal case of a pair of intersection on the wage curves $\omega(r)$ correspondingly to a pair of switching points in the r–p plane $p(r)$, if d were set $d = (0.95, 0.05)$ (Fig. 3.8).[10]

These kinds of example may be fabricated as you like. Thus there has seen a number of varieties on the side of $\omega(r)$, despite of facing to the real reswitching phenomena on the side of $p(r)$'s. This is the reason why we rather use the arguments of reswitching in the following by employing the r–p plane, instead of employing the wage curves.

9

$$w^{\alpha} = \frac{3.31181 - 1.32472\,r - 0.695479\,r^2}{3.28366 + r};\, w^{\beta} = \frac{-1.7601 + 2.20012\,r - 0.642436\,r^2}{-1.77506 + r}.$$

The points fulfilling $w^{\alpha} = w^{\beta}$ contain the complex conjugates:

$r_1 = -4.36542$, $r_2 = 0.47854 - 0.446038i$, $r_3 = 0.47854 + 0.446038i$.

10

$$w^{\alpha} = \frac{1 - 0.4\,r - 0.21\,r^2}{0.9905 + 0.30265\,r};\, w^{\beta} = \frac{1 - 1.25\,r - 0.365\,r^2}{1.0095 - 0.59405\,r}.$$

The points fulfilling $w^{\alpha} = w^{\beta}$ contain all the real values:

$r_1 = -4.11011, r_2 = 0.526316, r_3 = 0.614932.$

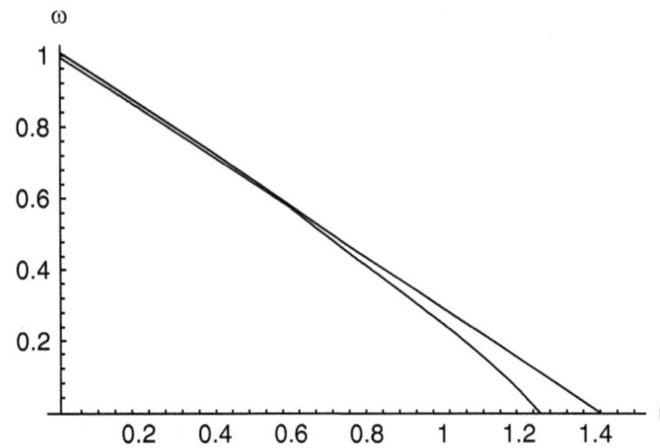

Fig. 3.7 A disguised example of non-reswitching given $d = (0.84, 0.15)$

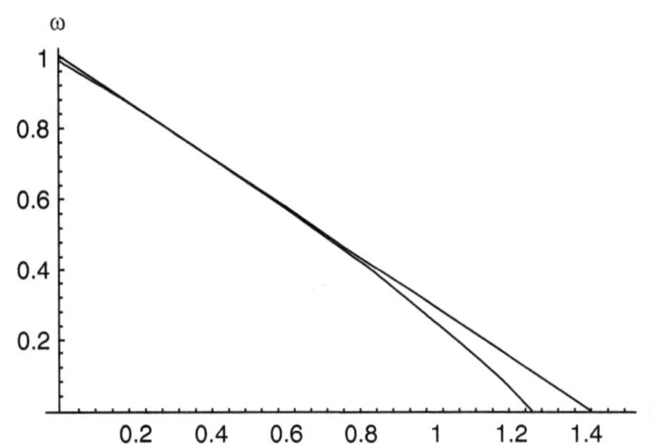

Fig. 3.8 A disguised example of non-reswitching given $d = (0.95, 0.05)$

3.2.2 The Sraffian Principle of Distribution

It however is natural that we should be true to our Sraffian principle when dealing with the Sraffian price system. This price system originally took off just by rejecting the neoclassical distributive principle. Our system must be subject to the following distributive principle:

$$(1+r)pA + wl = p,$$

as being associated with the solution of

$$p = [I - (1+r)A]^{-1}l,$$

which is uniquely nonnegative for $r < R$[the maximum r], if $I - A$ is nonnegatively invertible. Much token for reswitching were shown in the province of the neoclassical principle. On the other hand, to my experience on numerical simulation, we have never discovered any reswitching pair on the reasonable nonnegative rates of profit. It shall rather be curious that there has not been shown a thoroughgoing investigation for reswitching even in the Sraffa traditional two-sector model. In the next section, we can conclude that there never exists any pair of reswitching in the reasonable range over nonnegative rates of profit when we are engaged into the two-sector model of Sraffa principle of distribution.

3.2.3 Possibility of Reswitching in the Sraffa Two-Sector Model

Given a normalisation of labour vector being kept common among any two systems under observation, it is an easy task to prove impossibility of reswtiching in the Sraffa traditional two-sector model. We, first of all, prove impossibility in this case.

The two systems we take are different only from the first process of production while the second is common both in the two systems. The first process in a given system is written by (a_{11}, a_{12}); the first alternative one in a new technological system written by (a_{01}, a_{02}) as being discretely adjacent to (a_{11}, a_{12}). Thus a given system is represented as $[A, I, l]$, a new one as $[A^n, I, l^n]$, provided that $l = l^n = (1{,}1)'$. *The first normalisation hypothesis on labour* implies that input and output coefficients are measured in terms of labour input.

$$A = \begin{bmatrix} a_{11} & a_{12} \\ a_{21} & a_{22} \end{bmatrix}, \quad l = \begin{bmatrix} 1 \\ 1 \end{bmatrix}, \quad A^n = \begin{bmatrix} a_{01} & a_{02} \\ a_{21} & a_{22} \end{bmatrix}, \quad l^n = \begin{bmatrix} 1 \\ 1 \end{bmatrix}.$$

It is also assumed that nonnegative and indecomposable properties all are satisfied always in both systems. This can assure unique maximum nonnegative real-eigenvalues $\lambda(A)$, $\lambda(A^n)$ respectively associated with nonnegative price solution, thus establishing nonnegative maximum rate of profit R, R^n. In addition, it is supposed to be held net-producibility in a strict sense for each good:

$$a_{11}a_{22} - a_{12}a_{22} > 0, a_{11}a_{22} > 0, a_{12}a_{22} > 0\,,$$

$$a_{01}a_{22} - a_{02}a_{22} > 0, a_{01}a_{22} > 0, a_{02}a_{22} > 0\,.$$

Under these requirements, we obtain the unique nonnegative prices of the form:

$$p = [I - (1+r)A]^{-1}l,$$

$$p^n = [I - (1+r)A^n]^{-1}l^n.$$

The solution can be expressed in terms of *rational polynomial*. In details, the prices of good 1 are shown in the following manner:

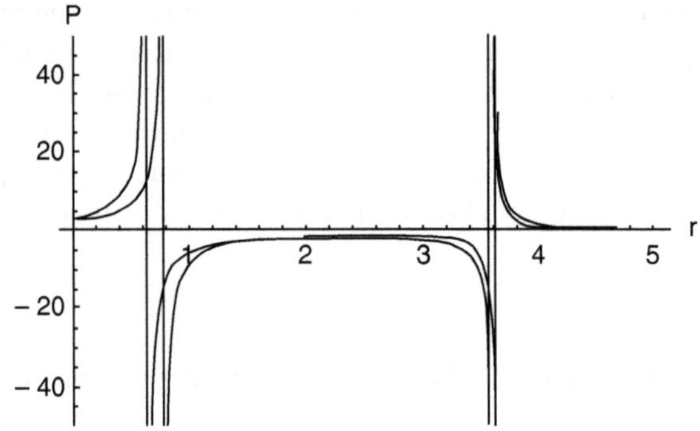

Fig. 3.9 Price behaviors on good 1 by two different techniques

$$p_1 = \frac{-1-a_{12}+a_{22}-a_{12}r+a_{22}r}{-1+a_{11}+a_{12}a_{21}+a_{22}-a_{11}a_{22}+a_{11}r+2a_{12}a_{21}r+a_{22}r-2a_{11}a_{22}r+a_{12}a_{21}r^2-a_{11}a_{22}r^2},$$

$$p^n{}_1 = \frac{-1-a_{02}+a_{22}-a_{02}r+a_{22}r}{-1+a_{01}+a_{02}a_{21}+a_{22}-a_{01}a_{22}+a_{01}r+2a_{02}a_{21}r+a_{22}r-2a_{01}a_{22}r+a_{02}a_{21}r^2-a_{01}a_{22}r^2}.$$

$p_1, p_1{}^n$ will be divergent, as r approaches R, R^n. These are the rising *hyperbolic curves* for $r \in [0, R)$, $[0, R^n)$. It is well-known in the Sraffian price system that $p_i(r)$ is positive for $r < R$. It must be also stressed in our context that $p_i(r)$ turns to be negative, if r exceeds R, since the price curve jumps the other hyperbolic curve at the asymptotic line R. Similarly as the same things applies to the alternative system (Fig. 3.9).[11]

Our task to look for reswitching is no less than to solve the roots of the equation:

$$y = p_1(r) - p^n{}_1(r) = 0.$$

A brief inspection of orders of the rational polynomials described in the above gives y a minimal polynomial of quadratic order. So, a minimal polynomial of y can by factorisation be the form:

$$(r+a)(r^2+br+c),$$

which in general gives three roots:

$$r = -a,$$

$$r = -b \pm \sqrt{b^2 - 4c/2}.$$

Leaving the signs of r aside, there may be three roots in total algebraically a possible pair of which can remain as reswtiching points.

[11]In the Sraffian price equation, 1/(1 + eignevalues) plays a roll of the asymptotic line. We have two eigenvalues derived from the two-sector model, excepting multiple roots. There usually are four asymptotic lines in total in both system under observation.

3.2.4 The First Proof of Possibility Theorem on Reswitching

A full calculation in the case of *the first normalisation hypothesis on labour* gives the following results:

$$r_1 = -1, r_2 = -1 + (1/a_{22}),$$

$$r_3 = \frac{a_{01} + a_{02} - a_{11} - a_{02}a_{11} - a_{12} + a_{01}a_{12} + a_{02}a_{21} - a_{12}a_{21} - a_{01}a_{22} + a_{11}a_{22}}{a_{02}a_{11} - a_{01}a_{12} - a_{02}a_{21} + a_{12}a_{21} + a_{01}a_{22} - a_{11}a_{22}}.$$

The first two of roots have definitely simple forms while the third one consists of quite complex combinations of a_{ij}'s, to which an economic implication can never be attached. $r_1 = -1$ is obtained form the normalisation hypothesis on labour. We however omit negative rates of profit since they are considered economically meaningless. One of three candidates has then been removed. Thus we could not assure two switch points on the reasonable range of $r \in [0, R)$, provided that $R \leq R^n$, if one of the remaining two should lose economic implication as a viable switch point. It may be evident by inspecting r_2 that it must be the case (Fig. 3.10).

The figure shows a *hypothetical* situation in which an economically viable reswitching happens to appear. By means of this figure, conversely, it turns out that a switch point r_2 may not be a viable point, if r_2 were placed on the right-hand side of the price curve associated with a given system whose maximum rate of profit is smaller than the other system's. Notice that prices must turn their sign into the opposite, if r exceeds a maximum rate R. We can in fact prove that the r_2 exceeds the asymptotic line of a given system.

We suppose to be $R \leq R^n$, without loss of generality. Since $\lambda(A) = 1/(1+R)$,

$$\lambda(A) > \lambda(A^n).$$

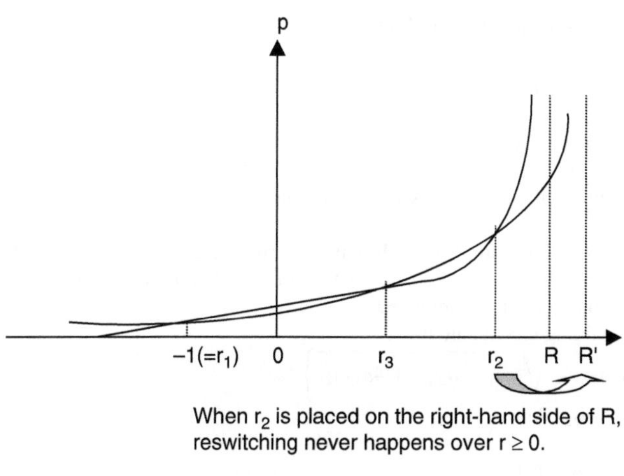

Fig. 3.10 A hypothetical situation for non-reswitching

On the other hand,

$$r_2 = -1 + (1/a_{22}).$$

In order to verify that R is placed on the right-hand side of r_2, it is convenient to transform the expression of r_2 into the similar form as $1/(1 + R)$:

$$1/(1 + r_2) = a_{22}.$$

Thus it is seen that

$$r_2 > R \Leftrightarrow a_{22} < \lambda(A).$$

We may easily check $a_{22} < \lambda(A)$. As the eigenvalues of matrix A are

$$\lambda = \frac{(a_{11} + a_{22}) \pm \sqrt{(a_{11} - a_{22})^2 + 4a_{12}a_{21}}}{2},$$

the maximum eigenvalue is

$$\lambda(A) = \frac{(a_{11} + a_{22}) + \sqrt{(a_{11} - a_{22})^2 + 4a_{12}a_{21}}}{2}.$$

Hence it follows that[12]

$$\lambda(A) > a_{22}.$$

[12]Reformulating this inequality, it follows

$$(a_{11} - a_{22}) + \sqrt{(a_{11} - a_{22})^2 + 4a_{12}a_{21}} > 0.$$

This will be satisfied even if

$h = a_{11} - a_{22} < 0$.

Since we have chosen the positive sign of $\sqrt{\ }$, it holds

$h \leq \sqrt{h^2}$

as well as $4a_{12}\,a_{21} > 0$, by supposition. It may be permissible that $4a_{12}\,a_{21} \geq 0$. The case of $4a_{12}\,a_{21}=0$ contains a possibility to give rise a switch point at the infinite rate of profit, which is made economically meaningless to be removed.

The other remaining eigenvalue is

$$\gamma = \left\{(a_{11} + a_{22}) - \sqrt{(a_{11} - a_{22})^2 + 4a_{12}a_{21}}\right\}/2.$$

This implies $\gamma < a_{22}$. Hence

$\lambda(A) > a_{22} > \gamma$.

Equivalently, it holds that

$R < (1 - a_{22})/a_{22} < (1/\gamma)-1$.

It is noted that the above argument is irrelevant to the previous condition of intensity on capital or labour in the first process of production. r_3 may remain the position on the left-hand side of R, of course. A switch point can viably exist for $r \in [0, R)$, but nothing for reswitching for the same range. Now we can summarise this as a proposition:

Proposition 2.1. *Input /or output coefficients are all measured in terms of labour input. We compare a given Sraffa two-sector production system with the alternative system which is technically different only from the first process. There exists no room for reswitching on the reasonable range of r, namely, r∈[0,R). Incidentally, a switch point may be permissible.*

3.2.5 A More General Proof with a Relaxation of Normalisation on Labour Input

In this section, a relaxation of *the first normalisation hypothesis on labour* will be introduced. We specify $l_0{}^n$, the labour input of the alternative process to produce good 1as z, while all the other labour inputs being fixed to be 1. The impossibility result on reswitching in the Sraffa two-sector model still holds by this kind of relaxation. Moreover, some interesting observations of bifurcation of root when being added some perturbation to the concerned system will be added. Replacing $l_0^n = 1$ with $l_0^n = z$, our production systems are of the form:

$$A = \begin{bmatrix} a_{11} & a_{12} \\ a_{21} & a_{22} \end{bmatrix}, \quad l = \begin{bmatrix} 1 \\ 1 \end{bmatrix}; \quad A^n = \begin{bmatrix} a_{01} & a_{02} \\ a_{21} & a_{22} \end{bmatrix}, \quad l^n = \begin{bmatrix} z \\ 1 \end{bmatrix}.$$

After solving $p = [I - (1+r)A]^{-1} l$, $p^n = [I - (1+r)A^n]^{-1} l^n$, we then solve $p_1 - p_1^n = 0$ respect with r. Thus we have the considerably more complex expressions than those in the previous case.

$$r_1 = -1 + (1/a_{22}),$$

$$r_2 = \left\{ (h - 2g) + \sqrt{(h+g)^2 - 4g(1 - (h+g))} \right\} / 2g,$$

$$r_3 = \left\{ (h - 2g) - \sqrt{(h+g)^2 - 4g(1 - (h+g))} \right\} / 2g.$$

Here

$$\begin{aligned} h &= a_{01} + (a_{02} - a_{12}) + a_{22} - a_{11}z - a_{22}z \\ &= (a_{02} - a_{12}) + (a_{01} - a_{11}z) + (a_{22} - a_{22}z), \end{aligned}$$

$$g = a_{02}a_{11} - a_{01}a_{12} - a_{02}a_{21} + a_{12}a_{21}z + a_{01}a_{22} - a_{11}a_{22}z$$
$$= (a_{02}a_{11} - a_{01}a_{12}) + (a_{01}a_{22} - a_{02}a_{21}) - (a_{11}a_{22} - a_{12}a_{21})z$$
$$= (a_{02}a_{11} - a_{01}a_{12}) + (a_{12}z - a_{02})a_{21} + (a_{01} - a_{11}z)a_{22}.$$

In front of these complex combinations of a_{ij}'s and z, we will only make a futile attempt to identify the signs of h, g by the use of the assumption of net-producibility, and by such an operation as

$$a_{01} > a_{11}, a_{02} < a_{12} (\rightarrow a_{01}a_{12} - a_{02}a_{11} > 0)[\text{vice versa}],$$

keeping capita or labour intensities in mind. We must classify the groups according to every sign of h and g in order to establish the size of r_2, r_3. We have already examined the property of $r_1 = -1 + (1/a_{22})$ which was the same as r_2 in the previous case. It has been proved that this is greater than a maximum rate of profit R of the given system, provided that $R \leq R^n$. In the present setting, we can get rid of r_1 from the candidates for reswitching. Our exercise therefore is to prove either that r_2, as sharing the square item $\sqrt{}$ with the sign $+$, is greater than R, or that r_3, as sharing the square item $\sqrt{}$ with the sign $-$, takes a negative value. We can prove the latter, indeed.

Theorem 2.1. *We arbitrarily assign any value to the labour input in the alternative process in order to relax the normalisation of labour inputs. Suppose also to be $R \leq R^n$, without loss of generality. It still holds not only that a switch point exceeds R, a smaller maximum rate of profit, but also that another switch point is placed in a strictly negative rate of profit. There is at most only one viable switch point in the reasonable range of rate of profit. If we compare the given Sraffa two-sector production system with an alternative system which is technically different only from the first process, there does not happen any viable reswitching. This result does not depend on the intensities condition of factor utilisation, again.*

Procedure of Proof. (See Appendix as for proof in details)
r_2, r_3 can be decomposed into $(h-2g)/2g$ and $\sqrt{(h+g)^2 - 4g(1-(h+g))}/2g$. If the first item,

$$(h-2g)/2g \leq 0,$$

and the second item

$$\sqrt{(h+g)^2 - 4g(1-(h+g))}/2g \text{ is positive},$$

r_3 is strictly negative,
since $(h-2g)/2g - \sqrt{(h+g)^2 - 4g(1-(h+g))}/2g$. If the first item however

$$(h-2g)/2g \geq 0,$$

r_3 may not be negative, unless

| the first item | < | the second item |.

r_2, r_3 commonly have denominator of $2g$. So $g = 0$ gives singular points. We thus investigate the case of $g \neq 0$.

3.2.6 A Numerical Analysis of Bifurcation Points

Theorem 3.1 has established that there is not any viable reswitching in the Sraffa two-sector model. The fact can be observed numerically. We start at an initial situation in which multiple roots happen and then revise the value of z. One candidate of our initial conditions is such as the situation:

$$a_{22} = a_{01} + a_{02} - a_{11}, z = 1.$$

In this case, $r = -1$ is a multiple root.[13] Figures 3.11–3.13 has depicted the three shots of branching process. As z varies, the multiple root breaks down. When $r = -1$ branches out, one accompanying the item of minus $\sqrt{}$ can be negative, the other accompanying plus $\sqrt{}$ may be not only positive but also often be repelled

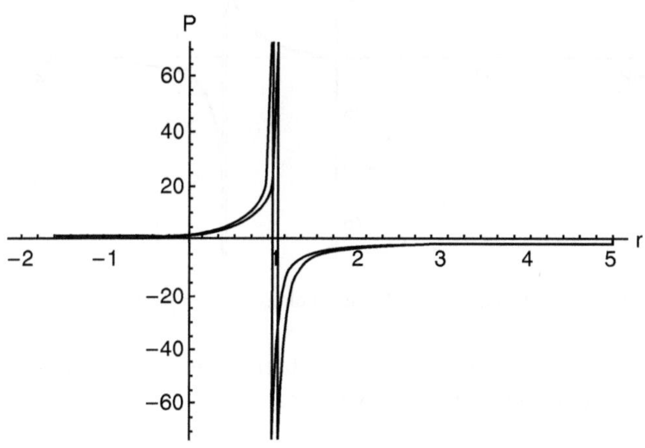

Fig. 3.11 Bifurcation in the case of $z = 1$

[13]If r_1 happens to be either r_2 or r_3, we may solve the equations $r_1 = r_2, r_1 = r_3$ with respect to a_{22}. The solutions are the same, irrespective of either $r_1 = r_2$ or $r_1 = r_3$:

$$a_{22} = \frac{a_{02}a_{11} - a_{01}a_{12} - a_{02}a_{21} + a_{01}a_{22}z}{a_{02} - a_{12}}.$$

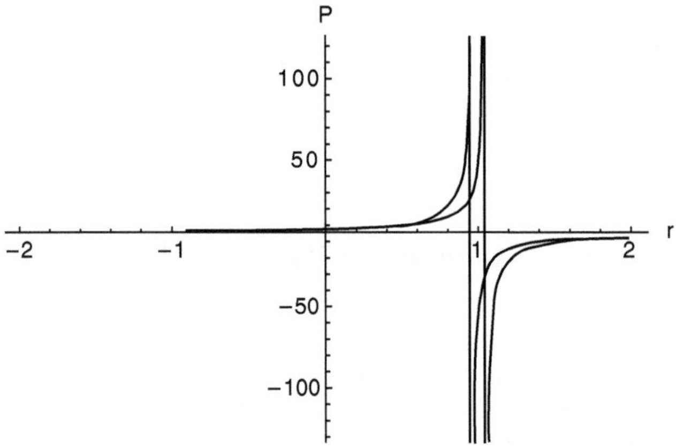

Fig. 3.12 Bifurcation in the case of $z = 0.95$

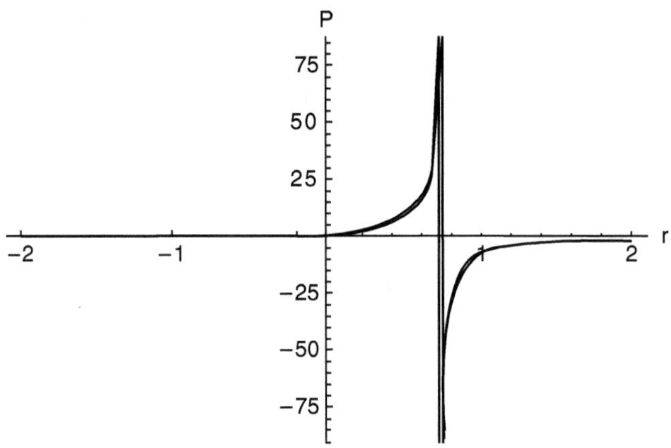

Fig. 3.13 Bifurcation in the case of another technology but $z = 0.95$

over $r_1 = (1 - a_{22})/a_{22}$. Thus we have observed that a switch point around a branching point never contributes reswitching.

3.2.7 A Numerical Exemplification of Reswitching in a Joint-Production System

A adoption of a Joint-production model even without the exhaustive distribution principle being imposed makes easier a probability of appearance of reswitching. We can immediately show this fact by the use of numerical example can be observed. It is noted that the first normalisation hypothesis on labour inputs is discarded here.

$$A = \begin{bmatrix} 0.45 & 0.15 \\ 0.15 & 0.38 \end{bmatrix}, \quad l = \begin{bmatrix} 1 \\ 1 \end{bmatrix}; \quad A^n = \begin{bmatrix} 0.4 & 0.2 \\ 0.15 & 0.38 \end{bmatrix}, \quad l^n = \begin{bmatrix} 0.95 \\ 1 \end{bmatrix}.$$

We, for convenience, set the output matrix as follows:

$$B = B^n = \begin{bmatrix} 1 & 1 \\ 1 & 1 \end{bmatrix}.$$

The three roots satisfying $p_1 = p^n{}_1$ are found to be

$$r_1 = 1, r_2 = 1.42563, r_3 = 1.63158.$$

It is seen that r_2, r_3 are a reswitching pair on an economically viable range (Figs. 3.14–3.15).

Why a joint-production can induce reswitching in a low dimension of production system seems closely related to the exhaustive distribution principle of factors income. To illustrate this, we examine the Sraffian joint-production system [A, B, l]. A price equations system of this joint-production system will be:

$$Bp = (1 + r)Ap + wl.$$

We set $w = 1$. We take an assumption on $[B–A]$ to be invertible.

$$p = [B - (1 + r)A]^{-1}l.$$

The solution can, by the use of the existence of $[B–A]^{-1}$, be transformed into the form:

$$p = \left[I - r[B - A]^{-1}A\right]^{-1}[B - A]^{-1}l.$$

The latter form of solution reflects the idea of a vertically integrated system by which the joint-production system can be reduced into a single-production system. We call the latter form the price system of a vertically integrated system:

$$\left[[B - A]^{-1}A, I, [B - A]^{-1}l\right].$$

In this bracket, the first component $[B–A]^{-1}A$ is supposed to be as input, the second I as output, and the third $[B–A]^{-1}l$ as labour. The elements of $[B–A]^{-1}A$ are the input coefficients required for the final goods of a vertically integrated system. The elements of $[B–A]^{-1}Al$ are the labour inputs indirectly required for an integrated system. The elements of $[B–A]^{-1}l$ are the direct labour inputs for an integrated system.

In this context, our production of goods is considered the reduced relation of input–output given by a series of reduced inputs tracing back to the primary factor as a whole.

We suppose to be $[B-A]^{-1}A > 0$. We also set:

$$Z_J(r) = [B - A]^{-1}A;\ l_J = [B - A]^{-1}l.$$

It then holds for an appropriate rate of profit

$$p(r) = l + rZ_J l_J + r^2 Z_J{}^2 l_J + \cdots + r^n Z_J{}^n l_J + \ldots > 0.$$

or equivalently,

$$p = [I - rZ_J(r)]^{-1} l_J.$$

Replacing $Z_J(r)$ with A, the latter form is no less than a solution of a single-production price system with the exhaustive distribution principle. In view of a vertically integrated system, the Sraffian joint-production system can be regarded as a single production system with the exhaustive distribution principle being imposed. Thus we have illuminated an idea of reswitching link between the exhaustive distribution principle and the Sraffian joint-production.

3.2.8 The Necessary and Sufficient Condition for Reswitching in the Sraffa Two-Sector Model

Adoption of the exhaustive distribution principle implies algebraically to extend the limits of maximum rate of profit. Put another way, we move from an economy that

$$\lambda(A) < 1/(1 + r),\ \text{namely}, R = (1/\lambda(A)) - 1 > r$$

to another economy

$$\lambda(\mathrm{A}) < 1/r, \text{namely}, R = 1/\lambda(A) > r$$

since either $p = [I - rA]^{-1}\, l$ or[14] $p = [I - rZ_J(r)]^{-1} l_J$.[14]

Since the value of R playing a roll of asymptotic line is larger when the exhaustive distribution principle are imposed, the price curves may have more gentle slopes by which curves would mutually be closer.[15] The tails of price curves will contain the broader positive range. Thus probability of reswitching are higher the closer the price curves are. This argument indicates why a joint-production

[14] We denote the maximum eigen value of $Z_J(\ r) = [B–A]^{-1}A$ by $\lambda(Z_J)$. We may regard $\lambda(Z_J)$ as $\lambda(A)$.

[15] $R = 1/\lambda(A)$ in the exhaustive distribution principle. On the other hand, $r = 1/a_{22}$ is a root. In effect, it is seen by the fact of $\lambda(A) > a_{22}$ that $R < 1/a_{22}$.

model has a high capability of generating reswitching. That is all, algebraically in a general Sraffa model. In the two-sector model, on the contrary, we can show that there exists a reswitching in the Sraffa two-sector model of basic goods as long as the exhaustive distribution principle is valid.

Proposition 2.2. *Suppose to be held* the first normalisation hypothesis on labour, *i.e., all the labour inputs being fixed 1. Then compare a given Sraffa two-sector model of basic goods with the other one as being different only from the first process of production in the former one. Given the exhaustive distribution principle in both systems, then, there remains some room for reswitching to happen.*

Proof. Given our presumptions, it follows:

$$p = [I - rA]^{-1} l,\ p^n = [I - rA^n]^{-1} l^n,\ l = l^n = (1, 1)'.$$

We can then solve $p_1 = p^n{}_1$ with respect to r.[16] The solution is as follows:

$$r_1 = 0, r_2 = 1/a_{22},$$

$$r_3 = \frac{a_{01} + a_{02} - a_{11} - a_{12}}{a_{02}a_{11} - a_{01}a_{12} - a_{02}a_{21} + a_{12}a_{21} + a_{01}a_{22} - a_{11}a_{22}}.$$

As for $r_2 = 1/a_{22}$, we can remove it as $R < 1/a_{22}$ in view of footnote 15. However, we cannot deny a possibility that it will belong to a range of $[0, R]$, since a property of r_3 is ambiguous. If $r_3 \in [0, R)$, there may be a reswitching pair of r_3 and $r_1 = 0$. □

Figures 3.14 and 3.15 show a numerical example of *Proposition* 2.2.

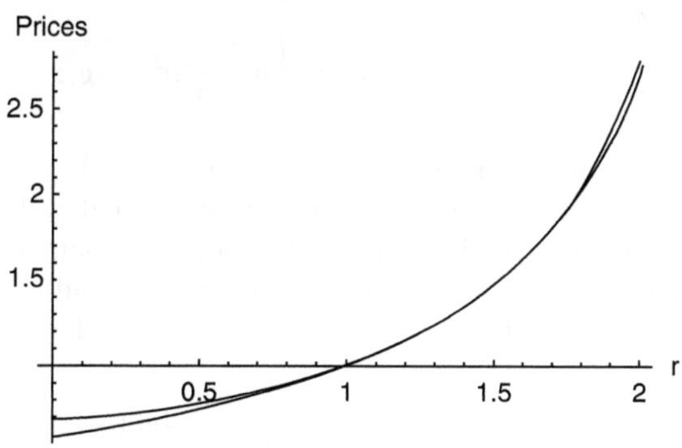

Fig. 3.14 A switch point at $r = 1$.

16

$$p_1 = \frac{-1 - a_{12}r + a_{22}r}{-1 + a_{11}r + a_{22}r + a_{12}a_{21}r^2 - a_{11}a_{22}r^2},$$

$$p^n{}_1 = \frac{-1 - a_{02}r + a_{22}r}{-1 + a_{01}r + a_{22}r + a_{02}a_{21}r^2 - a_{01}a_{22}r^2}.$$

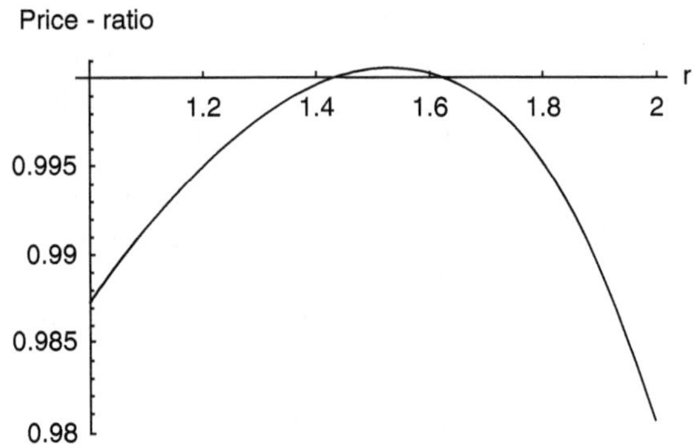

Fig. 3.15 A reswitching pair (Two price curves cross at two points, but they seem overlapped.)

Theorem 2.2. *The first normalisation hypothesis on labour is relaxed. We are discretionary to change a labour inputs of the first process of production in the alternative system. The two system under comparison are different only from the first process of production. Given the exhaustive distribution principle in both systems, even in this case, there can exist a reswitching pair both at strictly positive rates of profit.*

Proof. Setting $l^n = (z,1)'$, the roots are solved as follows:

$$r_1 = -1 + (1/a_{22}),$$

$$r_2 = \left\{h + \sqrt{h^2 - 4g(z-1)}\right\}/2g,$$

$$r_3 = \left\{h - \sqrt{h^2 - 4g(z-1)}\right\}/2g.$$

Compared these with the roots r_1, r_2, and r_3 in the previous theorem, r_1 is the same but the first items in the present r_2 and r_3 degenerate both into h, instead of $h{-}2g$. We can therefore save a somehow complicated classification of $h \geq 2g$. In the following, we assume $g \neq 0$ again to avoid infinite values of switch point.

We take the case of $h > 0$, $g > 0$. It then holds that, at r_2 and r_3,

$$h < \left|\sqrt{h^2 - 4g(1-z)}\right| \textit{ if } z > 1.$$

It however is fulfilled in the case of $z < 1$ that if

$$h^2 \geq 4g(1-z),$$

then

$$h > \left|\sqrt{h^2 - 4g(1-z)}\right| > 0 (g > 0).$$

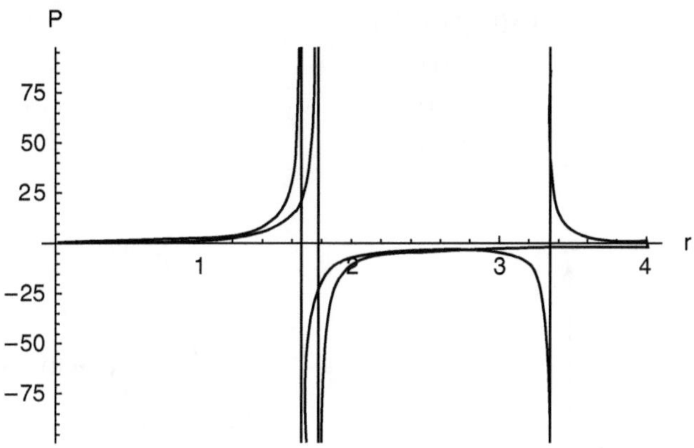

Fig. 3.16 A numerical illustration for Proposition 2.2

Fig. 3.17 A numerical illustration of the wage curve for Proposition 2.2

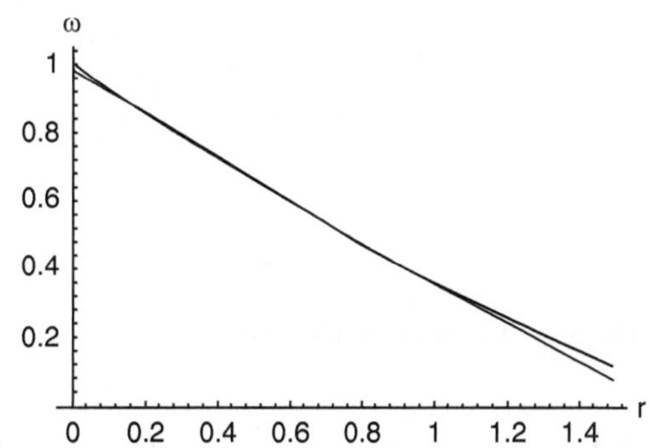

Hence, it can happen to be $r_3 > 0$. Thus there may be not only $r_2 > 0$ but also $r_3 > 0$, as indicating a possibility of viable reswitching on the strictly positive range as shown in Figs. 3.16 and 3.17. □

Corollary 2.1. *A necessary and sufficient condition for reswitching in the Sraffa two-sector model is the exhaustive distribution principle of factors income.*

Proof. Necessity) By *Theorem* 2.1

Invalidation of the exhaustive distribution principle of factors income ⇒ No reswitching on the viable range of rate of profit.

This gives the contraposition:

Reswitching ⇒ Validation of the exhaustive distribution principle of factors income

Sufficiency) By *Theorem* 2.2

The exhaustive distribution principle of factors income ⇒ Reswitching.

Hence the statement of the theorem has been proved. □

Figure 3.18 shows the image of Theorem 2.2.

$$p = [I - (1+r)A]^{-1} l$$
$$= M(r)\, l$$
$$l = (1, 1)$$
$$l^n = (1, z)$$

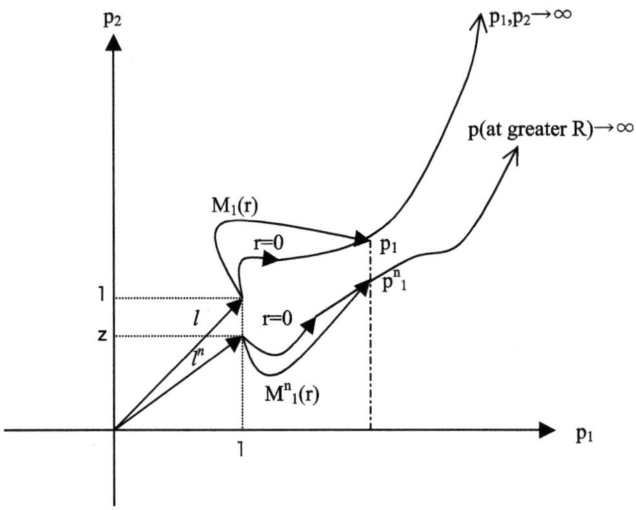

Fig. 3.18 The image of Theorem 2.2

3.3 A Short Discourse on Hyperbolic Linear Mapping and Reduction to Dated Quantities to Labour

It was noted in the end of Sect. 3.1 that a 3 by 3 Sraffian system rarely encounters a reswitching due to a strong hyperbolicity. We have a brief look on a recent development of hyperbolic mapping, since reswitching will be analysed with resort to convergence property of hyperbolic linear mapping.

3.3.1 Hyperbolic Linear Mapping

A recent study has developed a dynamical system on torus. In one-dimensional case of mapping, its periodic point is repeller unless it is attractor. A judgement whether it is repeller or not depends on sign of derivative. Eigenvalues derived from Derivatives of mapping determine whether a system is repelling or attracting. If the absolute value of eigenvalue were greater than 1, a system should be repelling. If the absolute value were less than 1, it should be attracting. This situation describes hyperbolic graph. The name of "hyperbolic" was adopted from this appearance.

3.3.1.1 Hyperbolic Mapping *H*

It is called "hyperbolic" if a vertible linear mapping does not have any absolute value of just 1 of eigenvalue, i.e., H is of automorphism

This mapping belongs to L, a linear mapping, and L_A, a linear mapping defined by matrix A.[17] A set of H_A is defined a subset of L:

$$H(R^{\mathrm{n}}) = \{H \in L(R^{\mathrm{n}})\}$$

It is verified that "H' still belongs to H (R^{n}) after perturbation is given H''. An extension of hyperbolic mapping H_A on torus will define periodic points under automorphism, one of the elementary concerns of nonlinear dynamical system. We however do not turn into a new frontier of mathematics, because mathematics at present is under analysing only in a lower dimensional system, and consequences available to us are not so many. We may only check a rudimentary fact of hyperbolic mapping for our purpose. Successive application of mapping L onto x constructs "a dynamical system":

$$\{L^0x, L^1x, L^2x, \ldots, L^nx, \ldots\}.$$

In a higher dimensional case, we have a mixed type of coexistence of repeller and attractor.

We can then distinguish the three cases to characterise periodic point in the following manner:

(1) All absolute values of eigenvalue is less than 1.
(2) All absolute values of eigenvalue is greater than 1.
(3) There are two distinct eigenvalue group whose absolute value is less than 1, on one hand, whose absolute value is greater than 1, on the other hand.

Proposition 3.1. *If it holds (1),* $H_A{}^n \to 0 (n \to \infty)$ for all $x \in R^n$;
If it holds (2), $H_A{}^n \to 0 (n \to -\infty)$ for all $x \in R^n$.

In order to promote our understanding, it is convenient to use a graphical presentation although its presentation is limited to a three-dimensional one. We can also

[17]Linear mapping is defined as follows: Let L be a linear mapping from $R^n \to R^n$. It is held that

$$L(u+v) = L(u) + L(v) \text{ for } u, v \in R^n; L(cu) = cL(u) \text{ for } (c \in R, u \in R^n).$$

Then L^{-1} also is a linear mapping. A linear mapping can be defined by a n by n matrix A. We denote this by L_A.

establish Proposition 3.2 as for a divergent trajectory of mixed type (3) in a three-dimensional case.

Proposition 3.2. *Let* λ *be eigenvalues of hyperbolic mapping* $H: R^3 \to R^3$. *Suppose that* $|\lambda_1| > 1, |\lambda_2| < 1, |\lambda_3| < 1$. *Then there are such a plane* W^s *and a plane* W^u *that*

(1) $x \in W^s \to H(x) \in W^s, H^n(x) \to 0(n \to \infty)$,
(2) $x \in W^u \to H(x) \in W^u, H^{-n}(x) \to 0(n \to -\infty)$,
(3) $x \in |W^s \cup W^u \to |H(x) \in W^s| \to \infty(n \to \infty)$.

3.3.2 The Sraffian Price System in Terms of Reduction to Dated Quantities to Labour

The arithmetic series of matrix form as expanded in the above linear parametric mapping *Z*(*r*), if we rather prefer a sum of cumulated series, can be easily interpreted with the Sraffian price system in terms of "reduction to dated quantities to labour". A solution of our price system, if *A* is defined by an invertible matrix *A*,

$$p = [I - (1 + r)A]^{-1} l$$

can be expanded into the following form:

$$p(r) = l + (1 + r)Al + (1 + r)^2 A^2 l + \cdots + (1 + r)^{\mathrm{n}} A^{\mathrm{n}} l + \cdots.$$

Here $A^n l$ is regarded as a dated labour thrown in the processes of the last preceding *n*-th period, according to the Standard Sraffian Theory. A definition of *Z*(*r*) gives another form of solution[18]:

$$p(r) = l + Zl + Zl^2 + \cdots + Z^n l + \cdots.$$

Hence our price solution is expressed in the form of the sum of cumulated linear mapping *Z*(*r*) onto a labour vector space *l*.

Hyperbolic *Z*(*r*) follows, if absolute eigenvalues of *Z*(*r*) are all less than 1. Invertible *Z*(*r*) follows from invertible *A*. This invertibility assures *Z*(*r*) to linear mapping for a particular *r*. Since a hyperbolic mapping *Z* onto *l* is convergent, the sum *p*(*r*) will be convergent, as shown in the last figures.

It is also defined for a similar hyperbolic mapping in a *Joint-production System* [*A*, *B*, *l*], if [*B*−*A*] is invertible

$$p = \left[I - r[B - A]^{-1} A\right]^{-1} [B - A]^{-1} l$$

[18] $(1 + r)^n A^n l = \{(1 + r)Al\} \ldots \{(1 + r)Al\} = \{Z(r)\} \ldots \{Z(r)\} = Z^n(r)$.

Here $[B-A]^{-1}A\, l$ is a labour input to produce goods of "a vertically integrated system" $[[B-A]^{-1}A, I, [B-A]^{-1}l]$. $[B-A]^{-1}\, l$ is a direct labour input for a vertically integrated system. We then reset $Z(r)$ as

$$Z_J(r) = r[B-A]^{-1}A\,.$$

We also reset l as

$$l_J = [B-A]^{-1}l\,.$$

Hence it follows the similar expansion:

$$p(r) = l + Z_J l_J + {Z_J}^2 l_J + \cdots + {Z_J}^n l_J + \cdots.$$

Our argument still holds in a joint-production system under the invertible assumption of $[B-A]$.

3.3.3 *The implication of a Parametric Matrix Z(r)*

A difference of prices $p(r) - p(r')$ has two alternative ways of illustration. One is a traditional illustration: its difference is a consequence of shift from mapping a particular r to another r' for a given system $[A, B, l]$. The other is an illustration by perturbation of technology: The original technology $(1 + r)A$ has changed into a new technology $(1 + r')A$. Or, it changed proportionally by $(r - r')A$. Thus we cannot distinguish the former from the latter. It turns out that it, in a certain case, is indifferent between a distributive change and a technological change.

A small change of r does not necessarily preserve hyperbolicity even if a linear mapping before undergoing a change is hyperbolic. We however know "continuity of eigenvalues" of linear mapping.[19]

[19]Continuity of eigenvalues is defined in the following way: Let $S(L)$ be a whole set of eigenvalues of L; λ to be any eigenvalue of L; m_λ to be multiplicity of $\lambda \in S(L)$; n to be the number of eigenvalues of a linear mapping L. It then holds:

$$\Sigma m_\lambda = n\,.$$

On the other hand, by introduction of another linear mapping T, suppose $\|T - L\| < \delta(\varepsilon)$ for a small $\varepsilon > 0$. We define $B_\varepsilon(\lambda)$ as a small ball with its origin λ (a respective eigenvalue of L) and its radius ε. Let λ' be eigenvalues of T. There may exist a small ball $B_\varepsilon(\lambda)$ in which $\{\lambda'_1, \ldots, \lambda'_k\}$, several kinds of eigenvalues of T, so to speak k, can be contained. We also define the intersection of this ball and S:

$$S(T) \cap B_\varepsilon(\lambda).$$

Here $\lambda' \in S(T) \cap B_\varepsilon(\lambda)$. In general, there may be multiple. We compare "λ in L" with "λ' in T". If the equality happened to be

$$\Sigma_{\lambda' \in S(T) \cap B\varepsilon(\lambda)} m_{\lambda'} = m_\lambda$$

for a respective λ, then we would call that eigenvalues are *continuous* in L.

Lemma 3.1 (Continuity of Eigenvalues in Linear Space). *Eigenvalues in $L(R^n)$ is continuously changeable by a perturbation of L.*

In the same spirit of *Lemma* 3.1, continuity of hyperbolic mapping can be proved:

Proposition 3.3 (Preservation of Hyperbolicity; Aoki (1996)Proposition 3.2.5). *$H(R^3)$ is an open set, and it is* dense *in $L(R^3)$.*

This proposition holds for a higher dimension. Resetting $\mu = -(1 + r)$ in the last proof, we can apply this to our hyperbolic mapping $Z(r)$. An appropriate selection of $\mu = -(1 + r)$, virtually r, always guarantees a different hyperbolic mapping $Z(r')$. For $r, r' \in [r_1, r_2]$, r' is taken in a small neighbourhood of r. Then $Z(r')$ will remain a similar configuration of eigenvalues, due to continuity. We show this kind of continuity by the use of a numerical example of 4 processes by 4 goods joint-production $[A, B, l]$ system used in the later section. A difference of Fig. 3.19 and Fig. 3.20 is due to a shift from matrix A to matrix $1.1A$ (A multiplied by 1.1) by an increase of scale factor from $r = 1$ to $r' = 1.1$ (Figs. 3.19–3.21).

A dynamical system generated by Z may however be divergent by the fact that $H(R^n)$ is open, on the other hand. In a special range $[r_1, r_2]$ there is a dense bundle of hyperbolic mapping Z whose destination may be convergent or divergent with or without various types of oscillation. *We restrict out interest to a convergent set of mapping.*[20]

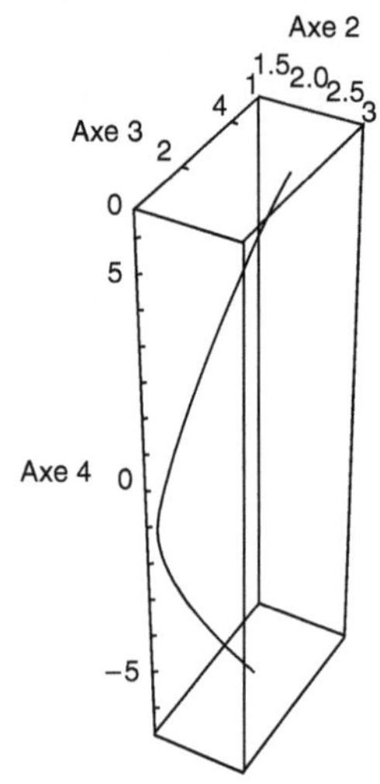

Fig. 3.19 A numerical example of the Parametric matrix $Z(r)$

[20]An arithmetic expansion of $Z(r) = r\,[B-A]^{-1}A$ is impossible, if a part of absolute eigenvalues of $Z(r)$ is greater than 1. It is noted that $r[B-A]^{-1}A$ can be convergently expansive for $r < 1$ even if $[B-A]^{-1}A$ is not. But $[I-Z(r)]$ in a certain case is nonnegatively invertible and then $p(r)$ has a nonnegative solution, even if it is the case.

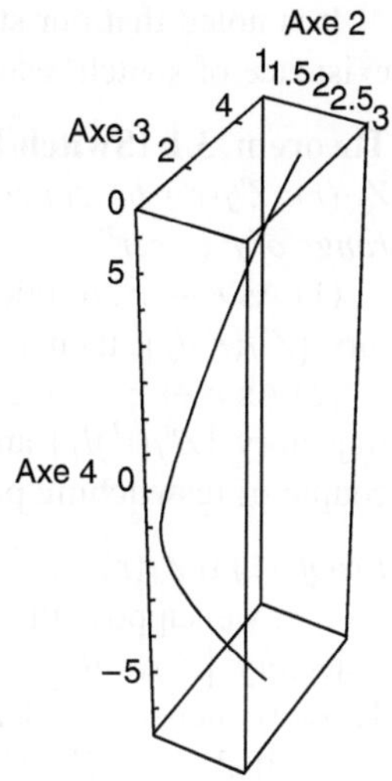

Fig. 3.20 A re-scaled form of the Parametric matrix $Z(r)$

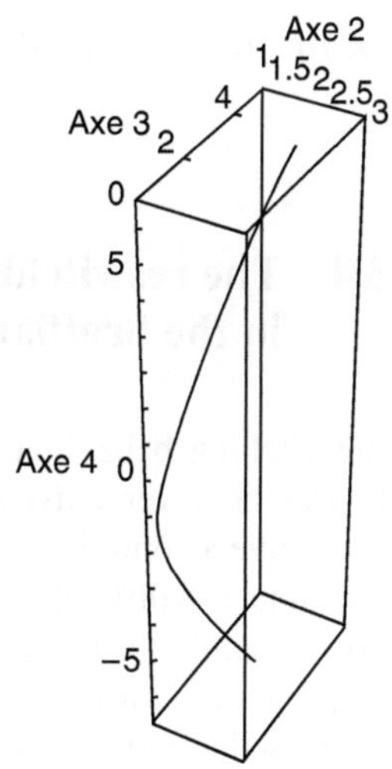

Fig. 3.21 Integration of Figs. 3.19 and 3.20

Next, we take any two hyperbolic mappings of different quality Z and Z' which both produce the convergent series. We can regard Z_J as based on $[B-A]^{-1}A$, while Z'_J as based on $[B'-A']^{-1}A'$.

$$p(r) = l + rZ_J l_J + r^2 {Z_J}^2 l_J + \cdots + r^n {Z_J}^n l_J + \cdots,$$

$$p'(r') = l + r' Z'_J l_J + r'^2 {Z'_J}^2 l_J + \cdots + r'^n {Z'_J}^n l_J + \cdots.$$

It is noted that our start point is the same initial condition l. Thus we can state existence of switch/reswitching points.

Theorem 3.1 (Switch/Reswitching Points Between Hyperbolic Systems). *Let $Z_J(r)$, $Z'_J(r')$ be convergent hyperbolic mappings in an appropriately specified range of $r \in [r, r'']$.*

(1) As $r \to r'$, a trajectory of a sequence $\{Z_J(r)l_J\}$ approaches toward a trajectory $\{Z'_J(r')l_J\}$, then a switch point exists.

(2) As $r \to r' \to r''$, a trajectory of a sequence $\{Z_J(r)l_J\}$ approaches toward a trajectory $\{Z'_J(r')l_J\}$ and approaches once more toward one of $\{Z'_J(r'')l_J\}$, then a couple of reswitching points exists.

Proof. (1) If $Z_J(r) = Z'_J(r')$ at $r = r'$ for each layer, then $p(r) = p'(r')$ at $r = r'$. As $r \to r'$, we suppose that a trajectory of a sequence $\{Z_J(r)l_J\}$ approaches toward a trajectory $\{Z'_J(r')l_J\}$ and passed it at $r = r'$. This situation defines a switch point. Those sequences are dense. Consequently, it must cross over it.

(2) If $Z_J(r) = Z'_J(r')$ at $r = r'$, as well as $Z_J(r) = Z'_J(r'')$ at $r = r''$ for each layer, then $p(r) = p'(r')$ at $r = r'$as well as $p(r) = p'(r'')$ at $r = r''$. As $r \to r' \to r''$, we suppose that a trajectory of a sequence $\{Z_J(r)l_J\}$ approaches toward a trajectory $\{Z'_J(r')l_J\}$ and passed it once at $r = r'$, and approaches once more toward one of $\{Z'_J(r'')l_J\}$ and passed it at $r = r''$. This situation defines a reswitching point. □

3.4 The reswitching Phenomena and Quasi-Attractors in the Sraffian Price System

Reswitching behaviour exceptionally seems perverse. As shown in the previous figures, however, a dynamical system generated by a hyperbolic mapping Z will by itself give a complicated trajectory in the linear space. Even if the initial points are the same, it is difficult to estimate where each convergent point arrives at. An orbit of prices seemingly appears *a wandering set*, indeed. *Randomness* will not be required to produce a twice-switch. We saw that the Sraffian price system $p(r)$ can be expressed in terms of a rational polynomial as stated in the end of Sect. 3.2. Hence the first and the second derivatives of $p(r)$ respect with r also are of the form of rational polynomial with higher degrees than the degrees of unprocessed prices. Thus it may be detected some definite relationship p–dp/dr–d^2p/dr^2, according to a method which are often used in the field of engineering when identifying chaotic property. This method is called *embedding*.[21] It is sure that the relationship p–dp/dr–d^2p/dr^2 can produce some orbits akin to chaotic attractor, as we call

[21]"Embedding" is a standard method to discover a causal nonlinear relationship embedded into data. This method simply constructs a phase diagram of Function–First Derivatives–Second Derivatives.

quasi-attractor. It must be noticed that our expressions of price do not have a counterpart of the difference equation forms, in despite of having the differential equation in Aruka (1991). Also noticed that r cannot be equivalently replaced with "time". These considerations may prevent our orbits from identifying with chaotic orbits. We may have the prediction:

Prediction 4.1. *Select a convergent series of mapping* Z *for an appropriate range. At a couple of reswitching points for* $r' \in [r, r'']$ *where a switch point comes out at least twice, the relationship* $p\text{–}dp/dr\text{–}d^2p/dr^2$ *gives a continuously nonlinear one like parabolic orbit or ellipse.*

3.4.1 A Numerical Example of the Joint-Production System

We now give a test in our numerical example. This will also be regarded as a test to reject a hypothesis that reswitching is a random phenomenon. As referred at the end of Sect. 3.2, a relative price in the 3 by 3 case will be reduced to essentially a rational quadratic form. It is pertinent for groping for a reswitching to use at least to 4 processes by 4 goods system, since hyperbolicity in the essentially over three-dimensional case will then contribute an existence proof for reswitching. Here we use an example of a 4 by 4 joint-production system. In this case, we can use a three-dimensional graphics by a selection of numéraire good to reduce one dimension. A 4 by 4 system is the highest one visually depicted. Even in such a case, it is difficult to find out a reswitching phenomena between hyperbolic systems. We rather prefer a reswitching between *subsets* of a system, namely, *intra-reswitching*.

3.4.1.1 Reswitching Between Subsets: Intra-Reswitching

If a price of process i at a particular r' happened to be equal to a price of another process j at the same r as a particular one for a given system, there would be called to be an intra-switch point between subsets $\{i\}$ and $\{j\}$ at the rate of r'. If the same thing could occur for another $r = r''$, there would be called to be an intra-reswitch point between subsets $\{i\}$ and $\{j\}$ at the rate of r''.

Let an input matrix A, a joint output B be as follows:

$$A = \begin{bmatrix} 0.1 & 0.4 & 0.1 & 0.2 \\ 0.3 & 0.2 & 0.1 & 0.1 \\ 0.1 & 0 & 0.3 & 0.1 \\ 0.1 & 0.09 & 0.6 & 0.38 \end{bmatrix}, \quad B = \begin{bmatrix} 1 & 0 & 0.15 & 0.3 \\ 0 & 1 & 0.2 & 0 \\ 0.15 & 0.2 & 1 & 0 \\ 0.2 & 1 & 0 & 0 \end{bmatrix}$$

Let a labour vector be:

$$l = (0.3, 0.4, 1.7, 0.4)'.$$

Since a price solution is

$$p = [I - r[B - A]^{-1}A]^{-1}[B - A]^{-1}l,$$

a hyperbolic mapping $Z(r)$ onto $[B{-}A]^{-1}l$ is of the form:

$$r[B - A]^{-1}A,$$

as previously considered.

We must, first of all, check whether absolute eigenvalues are less than 1 or not. In a joint-production case, we may check hyperbolicity for $r[B{-}A]^{-1}A$, as noted in footnote 20. We choose $r = 0.931$ as a value of r. Then it follows:

$$\lambda = \{0.779544 + 0.626092i, 0.779544 - 0.626092i, -0.127068, 0.0928181\},$$

$$|\lambda| = \{0.99984, 0.99984, 0.127068, 0.0928181\}.$$

If $r = 0.94$ is chosen, the first two items of $|\lambda|$ will exceed 1. In our example, a choice of r less than 0.931 will guarantee hyperboliciy.

Convergence of hyperbolic mapping for $r[B - A]^{-1} A$ around $r = 0.65$ shows spiral. The form of spiral is unchanged by a perturbation of r. We make a shift of r from $r = 0.65$ to 0.69 a point by point. In order to make "our convergent mapping of 4 by 4 matrix" visual, we select the first good as numéraire good. In other word, we divide each price by the price of the first good p_1. Trajectories of convergence are depicted by p_2/p_1, p_3/p_1, p_4/p_1 coordinates as the first, the second and the third axes (Fig. 3.22).

The solution surely is of the form of rational polynomial as already pointed out:

$$p_1 = 2(1290r^3 - 30967r^2 - 33020r + 17777)/\,(1504r^4 + 3830r^3 - 75961r^2 - 21060r + 46367),$$

$$p_2 = 4(-3306r^3 - 9681r^2 - 6874r - 379)/(1504r^4 + 3830r^3 - 75961r^2 - 21060r + 46367),$$

$$p_3 = 2(-10210r^3 - 71313r^2 - 9940r + 39103)/(1504r^4 + 3830r^3 - 75961r^2 - 21060r + 46367),$$

$$p_4 = 4(8278r^3 + 55873r^2 + 21682r - 56513)/(1504r^4 + 3830r^3 - 75961r^2 - 21060r + 46367).$$

These equations give a diagram of $r - p(r)$ plot (Fig. 3.23):

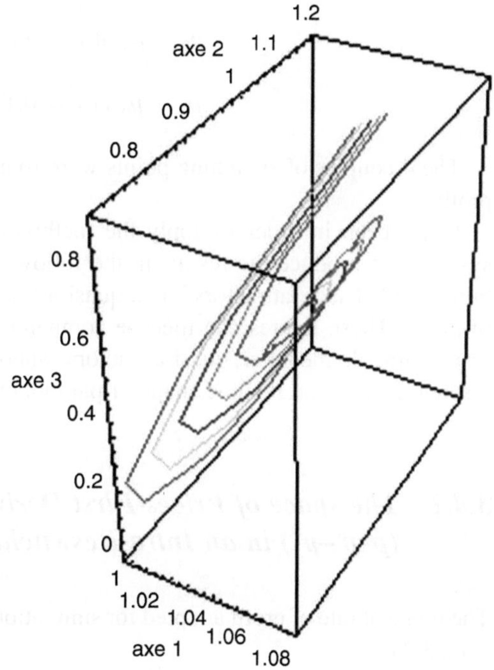

Fig. 3.22 A numerical example for convergence of hyperbolic mapping

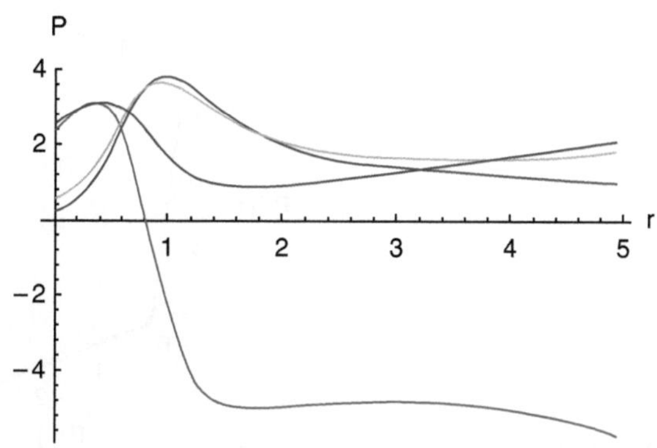

Fig. 3.23 The occurrence of reswitching in Fig. 3.22

The reswitch points between subsets are confirmed from this diagram:

$$p_1 = p_2 \text{ at } r = 0.78, 1.89\,,$$

$$p_1 = p_3 \text{ at } r = 0.69, 3.21,$$

$$p_1 = p_4 \text{ at } r = 0.58,$$

$$p_2 = p_3 \text{ at } r = 0.64, 4.17\,,$$

$$p_3 = p_4 \text{ at } r = 0.17, 0.40\,.$$

The 4 couples of switching points were found in the concerned range of rate of profit.

It has been in order to apply the method of embedding to the Sraffian price system. We produce the results in the following. We were able to observe something like "chaotic attractors", i.e., quasi-attractors in our concerned region of rate of profit. These figures can then be compared with those in the case of no intra-reswitching. In the latter, quasi-attractors cannot be produced. Trajectories could be discontinuous even if some shape of parabola were obtained.

3.4.2 *The space of Prices-First Derivatives-Second Derivatives (p–p'–p'') in an Intra-Reswitching Case*

The range of rate of profit adopted for simulation is from $r = 0$ to $r = 10$: $r \in [0, 10]$ (Fig. 3.24).

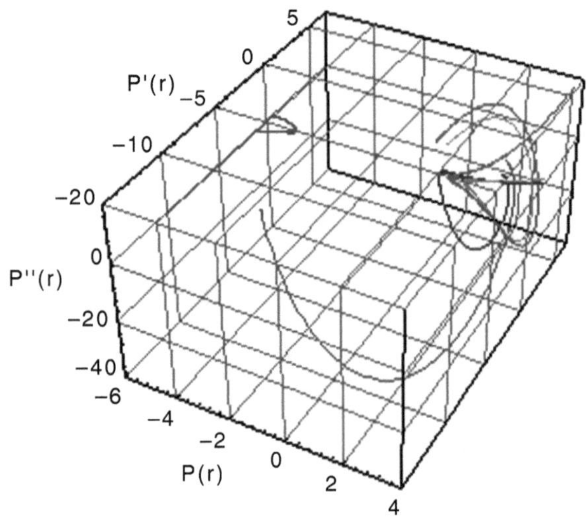

Fig. 3.24 A numerical example of intra-reswitching

3.4.3 The planes of Prices-First Derivatives in an Intra-Reswitching Case

Fig. 3.25

3.4.4 The planes of First Derivatives-Second Derivatives in a Reswitching Case

Fig. 3.26

3.4.5 A Numerical Example for No Intra-Reswitching Between Subsets

In order to show a case of no intra-reswitching, we only alter a system front a joint-production into a no-joint-production underlying the exactly same matrix A and the same labour input l:

$$A = \begin{bmatrix} 0.1 & 0.4 & 0.1 & 0.2 \\ 0.3 & 0.2 & 0.1 & 0.1 \\ 0.1 & 0 & 0.3 & 0.1 \\ 0.1 & 0.09 & 0.6 & 0.38 \end{bmatrix}, \quad B = I = \begin{bmatrix} 1 & 0 & 0 & 0 \\ 0 & 1 & 0 & 0 \\ 0 & 0 & 1 & 0 \\ 0 & 0 & 0 & 1 \end{bmatrix},$$

$$l = (0.3,\ 0.4,\ 1.7,\ 0.4)'.$$

3.4.6 The r–p(r) Plot in a No Intra-Switch/-Reswitching Case

Fig. 3.27

3.4.7 The space of Prices-First Derivatives-Second Derivatives (p–p′–p″) in a No Intra-Switch/Reswitching Case

The *range* of rate of profit adopted for simulation is from $r = 0$ to $r = 3$: $r \in [0, 3]$ (Fig. 3.28).

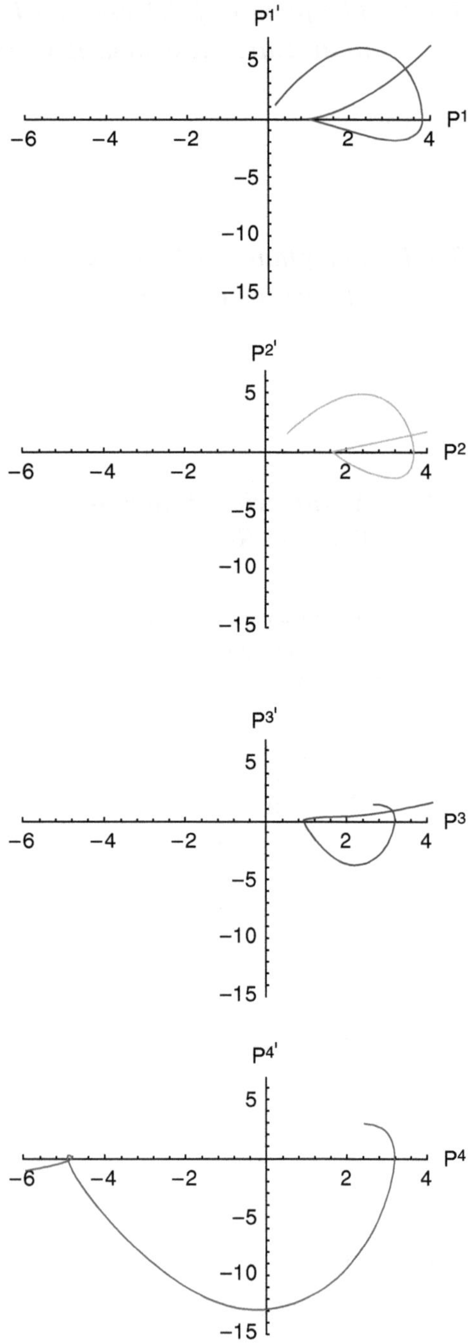

Fig. 3.25 The planes of prices-first derivatives in an intra-reswitching

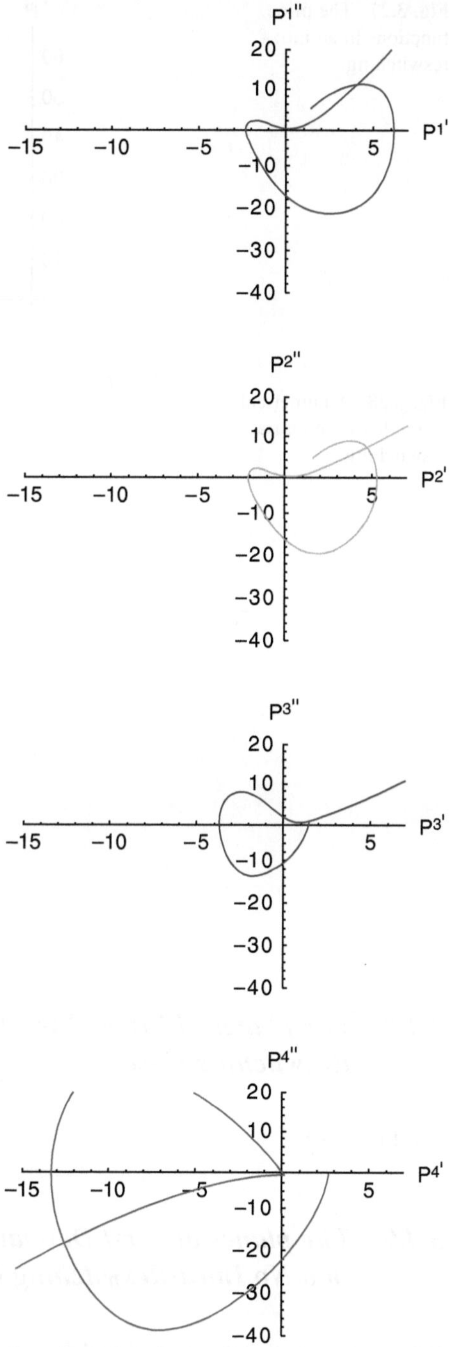

Fig. 3.26 The planes of first derivatives-second derivatives in an intra-reswitching

Fig. 3.27 The price functions in an intra-reswitching

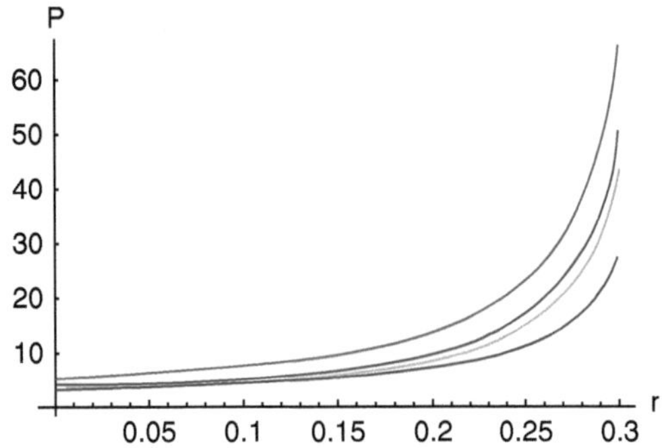

Fig. 3.28 A numerical example of a no intra-reswitching

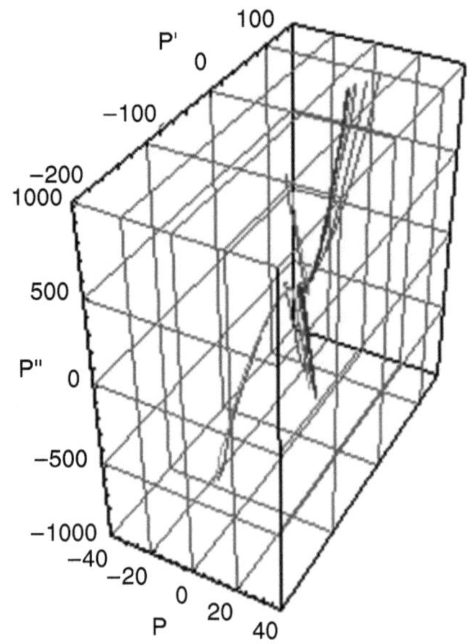

3.4.8 *The planes of Prices-First Derivatives in a No Intra-Switch/Reswitching Case*

See Fig. 3.29.

3.4.9 *The planes of First Derivatives-Second Derivatives in a No Intra-Reswitching Case*

It has turned out that there was found discontinuity in every plane in our case of no intra-switch/reswitching case. Thus our prediction as stated in the beginning, namely, *Prediction* 4.1 seems fulfilled (Fig. 3.30).

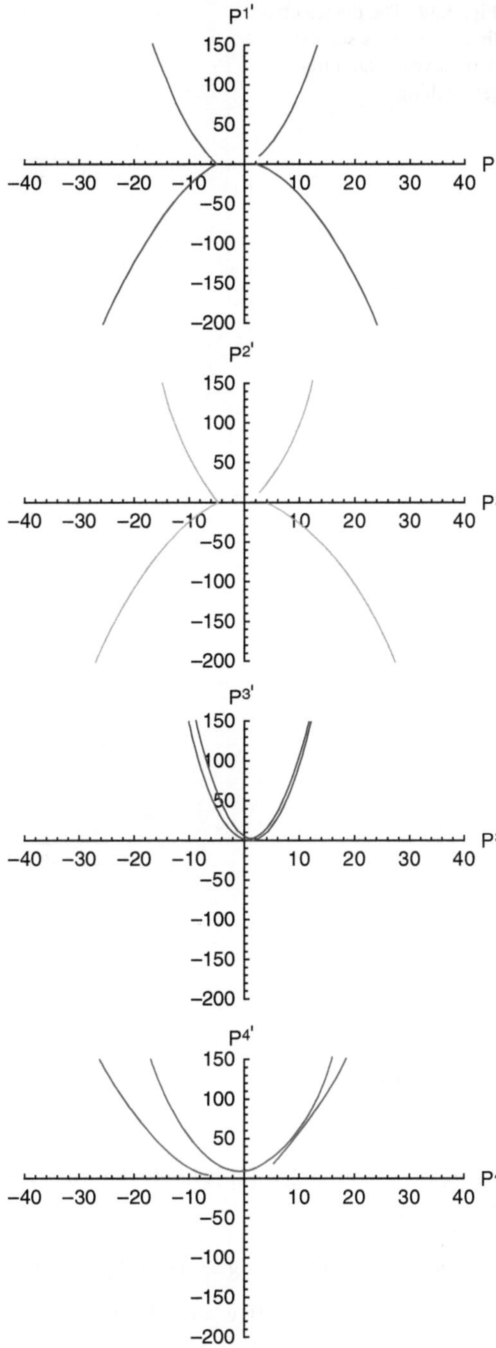

Fig. 3.29 The planes of prices-first derivatives in a no intra-reswitching

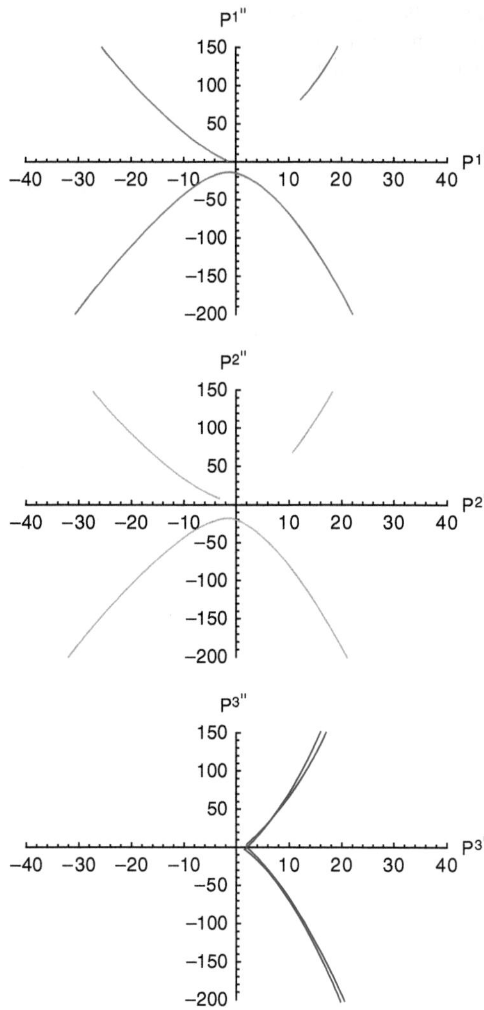

Fig. 3.30 The planes of first derivatives-second derivatives in an intra-reswitching

Appendix

Proof of Theorem 2.1. [**I**] It holds the following relations for any z.
Case (1) $h \leq 2g$:

$$h<0,\ g<0:\ h\leq 2g \rightarrow (h-2g)/2g=(h/2g)-1\geq 0, \text{ since } |h|\geq|2g|.$$

$$h>0, g>0: 0<h\ \leq 2g \rightarrow (h-2g)/2g\leq 0$$

$$h<0, g>0: h<0\ \leq 2g \rightarrow (h-2g)/2g<-1$$

$$h>0 \text{ and } g<0 \text{ contradicts } h\leq 2g.$$

Case (2) $h \geq 2g$:

$$h > 0,\ g > 0 : h \geq 2g \rightarrow (h-2g)/2g = (h/2g) - 1 \geq 0$$

$$h < 0, g < 0 : h \geq 2g \rightarrow (h-2g)/2g \leq 0, \text{since } |h| \leq |2g|.$$

$$h > 0, g < 0 : h > 0 \geq 2g \rightarrow (h-2g)/2g < -1$$

$$h < 0 \text{ and } g > 0 \text{ contradicts } h \geq 2g.$$

Thus it has seen for a nonnegative r_3 that
Case (3) as $h \leq 2g, h < 0, g < 0$, and
Case (4) as $h \geq 2g, h > 0, g > 0$.
We then examine Cases (3, 4).
Case (3) $h \leq 2g$ and $h < 0, g < 0 \rightarrow |h| > |g| > 0$:
Since $h < 0$, $g < 0$, in terms of absolute values,

$$|h + 2g| < \left|\sqrt{(h+g)^2 + 4g + (h+g))}\right|.$$

This inequality can be easily checked. Taking square of both hand sides and rearranging,

$$h^2 + 2hg + g^2 + 4g \geq h^2.$$

By our supposition of $|h| > |g| > 0$,

$$h^2 + 2hg + g^2 + 4g \geq h^2 > 0.$$

the numerator of r will be negative, given the sign of the second item being negative. It must be noted that the numerator is divided by $g < 0$ after it has been established

$$|h - 2g| < \left|\sqrt{(h+g)^2 - 4g(1-(h+g))}\right|.$$

In effect, r_2 turns to be negative, instead of r_3.
Case (4) $h \geq 2g$ and $h > 0$, $g > 0$:
We can check the inequality:

$$h - 2g < \sqrt{(h+g)^2 - 4g(1-(h+g))}.$$

Taking square of both-hand sides and rearranging[22],

$$(h+g)^2 + 8hg \geq 4g + h^{2\ 21}.$$

Remarking the relation

$$h^2 \geq (2g)^2, \text{namely}, h^2 \geq 4g^2,$$

it is seen that

$$(h+g)^2 + 8hg \geq 4g + h^2 \geq 4g^2 > 0.$$

Hence r_3 will be negative, since $g > 0$.

[II] We examine the case of multiple roots. We have already mentioned that a quadratic equation on r has a factorisation into primary factors of the form $(r + a)$ $(r^2 + br + c)$. Thus the roots of our problem is to be linked to the following correspondence:

$$r = -a \Leftrightarrow r_1 = -1 + (1/a_{22})$$

$$r = -b \pm \sqrt{b^2 - 4c/2} \Leftrightarrow r_2, r_3 = \left\{(h-2g) \pm \sqrt{(h+g)^2 - 4g(1-(h+g))}\right\}/2g$$

Hence the multiple roots happen in the form of $r_2 = r_3$ while it still remains $r_1 = -1 + (1/a_{22})$. The latter, however, is to be eliminated as a switch point due to its negativity by *Proposition* 2.1. Thus our supposed reasonable switch point will be only one, even if $r_2 = r_3$ were economically viable. In the case of $r_2 = r_3$, there is not any reswitching despite of formally two roots in the reasonable range. There will also be not any reswtiching still again in the other cases of multiple roots that $r_1 = r_2 = -1 + (1/a_{22}), r_1 = r_3 = -1 + (1/a_{22})$. We have proved that all the cases of multiple roots fail to give reswitching.[23]

Both results **[I]** and **[II]** assure the statements of our theorem. □

---

[22]Squaring both-hand sides

$$h + 2g < \sqrt{(h+g)^2 - 4g(1-(h+g))},$$

it follows that

$$g^2 + 10hg - 4g \geq 0.$$

Rearranging this, it will be given:

$$(h+g)^2 - h^2 + 8gh \geq 4g \rightarrow (h+g)^2 + 8hg \geq 4g + h^2.$$

[23]Our story hinges on the assumption that $l_0^n = z$, the others of labour inputs all being set 1. If one of the other labour inputs were relaxed to be another value than 1, our conclusion might not be much influenced. For instance, we set $l_1 = z_1$. Thus the definition of h as well as g may be extended to include z_1. The distribution of roots which can define switch points will remain formally similar as the previous case. The similar proof seems possible to give a similar result.

References

Aoki N (1996) Dynamical systems and chaos. Kyoritsu Shuppan, Tokyo (in Japanese)
Aruka Y (1991) Generalized Goodwin's theorems on general coordinates. Struct Change Econ Dyn 2:69–91. Reprinted in J. C. Wood (ed.) (1995) Piero Sraffa: critical assessments, vol. 3, Routledge, London, pp 127–153
Diewert WE (1982) Duality approaches to microeconomic theory. In: Arrow KJ, Intriligator MD (eds) Handbook of mathematical economics, vol 2. North-Holland, Amsterdam, pp 535–599
Fisher MF, Shell K (1972) The economic theory of price index: two essays on the effects of tastes, quality, and technological change. Academic, New York
Fisher MF, Shell K (1998) Economic analysis of production price index. Cambridge University Press, Cambridge
Levahri D, Samuelson PA (1966) A reswitching theorem is false. Q J Econ 80:68–76
Nelson RR (1996) The source of economic growth. Harvard University Press, Cambridge, MA
Newman P (1987) Duality. In: Eatewell J, Milgate M, Newman P (eds) New Palgrave: a dictionary of economics, vol 1. Macmillan and Maruzen, London, Tokyo, pp 924–934
Pasinetti L (1977) Lectures on the theory of production. Macmillan, London
Pasinetti L (1981) Structural change and economic growth. Cambridge University Press, Cambridge
Samuelson PA (1966) Summing up. Q J Econ 80:568–583
Schefold B (1989) Mr. Sraffa on joint-production and other essays. Unwin Hyman, London
Sraffa P (1960) Production of commodities by means of commodities. Cambridge University Press, Cambridge
Woodland A (1982) International trade and resource allocation. North-Holland, Amsterdam

Chapter 4
An Evolutionary Theory of Economic Interaction: Introduction to Socio- and Econo-Physics

Yuji Aruka and Jürgen Mimkes

Abstract We try to construct an evolutionary theory of economic and social interaction of heterogeneous agents. Modern physics is helpful for such an attempt, as the recent flourishing of econophysics exemplifies. In this article, we are interested in a specific or more fundamental use of physics rather than in the recent researches of econophysics. In the first part of this article, we mainly focus on the traditional von Neumann-Sraffa model of production as complex adaptive system and examine the measure of complexity on this model in view of thermodynamical ideas. We then suggest the idea of hierarchical inclusion on this model to define complexity of production. In the latter part, we try to construct an elementary theory of social interaction of heterogeneous agents in view of statistical mechanics.

Keywords Complex adaptive system · Complexity of production · Heterogeneous interaction · Hierarchical inclusion · Social temperature · Thermodynamics · Truncation of a production system

4.1 The Scenarios on Evolution and Selection

4.1.1 The Stories of Evolution

The society and economy can evolve. When he argued about what Evolutionary Economics was, Hodgson (2003) suggested that there was a priori a set of possible variations for selection of species in Darwinism, what we call an *ex ante* variation.

Reprinted from Evolutionary and Institutional Economics Review 2(2). Aruka, Y., An Evolutionary Theory of Economic Interaction: Introduction to Socio- and Econo-physics, 145–160 (2007). With kind permission from Japan Association for Evolutionary Economics, Tokyo.

Y. Aruka (✉)
Faculty of Commerce, Chuo University, Higashinakano, Hachioji, Tokyo 192–0393, Japan
e-mail: aruka@tamacc.chuo-u.ac.jp

J. Mimkes
Department of Physics, University of Paderborn, Paderborn, Germany
e-mail: mimkes@zitmail.uni-paderborn.de

Y. Aruka (ed.), *Complexities of Production and Interacting Human Behaviour*, DOI 10.1007/978-3-7908-2618-0_4, © Springer-Verlag Berlin Heidelberg 2011

Following Hodgson, here, we should not demonstrate unilaterally that a certain variation could emerge *ex post* as a result of selection process on species or social subgroups. Put another way, a newly arriving variation will emerge as the result of interacting variations. Hodgson thus thinks that there unambiguously exists a certain demarcation between Darwinism and Schumpeterian Economics, because Schumpeterian economics gives importance rather to resultant variation.

Sometimes we have the idea that the agent who happened to survive may be proven to be *ex post* superior. Even given superficially homogeneous agents, indeed, we could verify the emergence of *mutants*. This is just a Spencerian story of individualistic agents, as Keynes (1972, p. 28) heavily satirized. One that can perform better by itself than the others can survive to be *mutant*. This idea may then be reinforced in association with utility maximizing principle of individualistic homogeneous agents. The importance of marginal productivity in this context is just equivalent to the story of *giraffe*: The giraffe, with its higher neck, has the advantage of being to eat more leaves can thus survive more easily. Without heterogeneity, thus, we are sometimes used to seeing the occurrence of mutants.

4.1.2 The Internal Process of selection: Ex Ante Variations, Diversities, and Niches of Species or Social Subgroups

Suppose on the other hand that there were *ex ante* many heterogeneous agents. In this case, we have a new utility theory of interaction coming from multinomial logit model, even if we leave the domain of utilitarianism.[1] The idea of heterogeneous agents is not necessarily inconsistent with the utilitarianism. Heterogeneous interacting agents lead to bring a new set of variations, anyway.

We thus have *heterogeneous interacting agents* sometimes with bounded rationality, while, *homogeneous individualistic agents* always with complete rationality.[2]

In the process of selection faced to diversity, we will need to comprehend an internal process to delete a niche and add to a new agent, to more or less restore the missing interaction. Holland, who suggested the framework of *complex adaptive system*, described as follows:

> The diversity is neither accidental nor random. The persistence of any individual agent, whether organism, neuron, or firm, depends on the context provided by the other agents. Roughly, each kind of agent fills a niche that is defined by the interactions centering on that agent. If we remove one kind of agent from the system, creating a "hole," the system typically responds with a cascade of adaptations resulting in a new agent that "fills the hole". The new agent typically occupies the same niche as the deleted agent and provides most of the missing interactions (Holland 1995, p. 27)

[1]The multinomial logit model can be compatible with the sequential choice model, as Luce model (Luce 1959) suggested. The multinomial logit model has a comprehensively generic feature either to be derived from the random utility model or the Luce model. See Bierlaire (1997) for details.

[2]Bounded rationality may be given in view of the failure of estimation on the secondary effects. If we had some subpopulation which failed to catch up with the secondary effects of the interaction, the opponent subpopulation should rather utilize the failure of the former for its advantages.

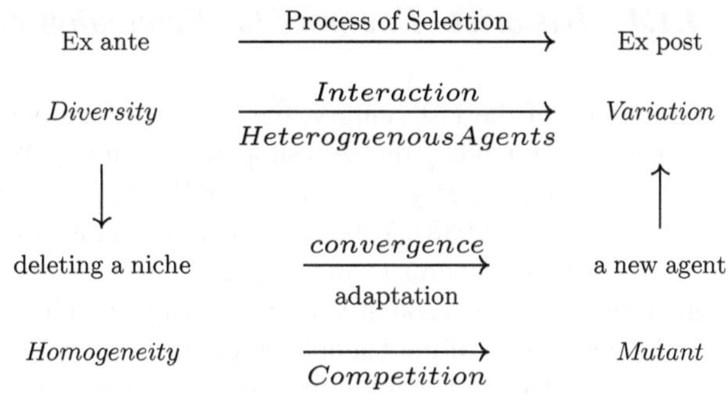

Fig. 4.1 A scheme of selection and variations

The visual formula of our main points may be summarized in Fig. 4.1:

4.1.3 Rules Choice

Principle in the above may be interchangeable with *rule*. Maximizing principle can be read as maximizing rule, for instance. Rule choice in economics from the first has been predetermined for the last decades. There then is no other room than the utility maximization rule in orthodox economics. If we should accept this, our research focus should be enclosed within the limited world of Descartes, where realism of material substance and spiritual agent are completely separated. *Empiricism* thus was inclined to be almost entirely removed from economic theory. Games of purely mathematical modeling then continued to be played long.

We know that aggregation of microscopic agents by itself implies the emergence of a new property which never appears in a state of a single agent. This is a true meaning of aggregation. We can, however, notice that our attitude to study Macroscopic Aggregate Production Function was quite different from this way. Hoover (2001, pp. 74–85) discussed such an attitude in details. We can also wonder to ourselves:

> Why did economists have had much fun to cling to prove a set of properties at the macroscopic level for preserving the microscopic properties of production? This is an inexplicable question in the traditional economics. In the end of the last century, however, this trend just began to be checked by a new stream from Physics. This stream symbolically means the rise of *econophysics*.

In orthodox economics, we only are permitted to use the sole rule, i.e., *utility* maximization as the *generic* principle. Utility must, however, be a partial principle. We must stop applying a straitjacket, because we will require another principle such as *risk* minimization when we are faced with uncertainty. Diverse rule choice should be allowed us.[3]

[3]*The multi-armed bandit problem* (See Bellman 1961), as Holland (1992, Chap. 5) referred to is a good example that optimization should fail. Our permissible option in such an environment that the problem is faced with is limited to the way of risk minimization.

4.1.4 Historical Excurses for Econophysics

At the end of the last century, the Santa Fe Institute has been quite well-known worldwide for holding the workshop in September 1987 titled: "Evolutionary Paths of the Global Economy" generated by P.W. Anderson, K.J. Arrow, and David Pines (Anderson et al. 1988). This conference was a historically monumental meeting to bring together economists and physicists. The success of this meeting has spurred the research on the economy as an evolving complex system.

Independently of the Santa Fe attempts, however in Stuttgart, Germany, it must be noticed that we had a big bang of the new approach to social phenomena. This approach was born in the process of Synergetics Project, stimulated, in particular, by Herman Haken's study on laser beam. This group calls its own approach Sociodynamics. Wolfgang Weidlich has played the decisive role on *the Sociodynamics Project*, which was compiled as Weidlich (2000).

Interestingly, this nomination coincides with the final layer of Schumpeterian economics, since we have the triangular theoretical layers of statics, dynamics, and socio dynamics as the stages of economic epistemology. Evolutionary economics may thus easily accept the study of sociodynamics, while the synergetics approach may accept evolutionary economics if it becomes engaged in any integration of physics and economics.

We can normally distinguish the variables of a dynamical system between *slow* variables and *fast* variables. By referring to slow variables, the Synergetics approach in brief is used to find *the order parameter* in order to construct *the master equation* for studying the dynamical properties of a system. The sociodynamics approach thus introduced the ideas of a master equation into social and economic analysis. In other words, the idea of statistical mechanics could, together with the master equation, be applied to social science. This approach never depends on the so-called classical mechanics. In the new approach, each state could be described in *a detailed balance* of in-flow and out-flow, and the emergence of *fluctuations* of state can also be analyzed.

4.2 Entropy and Complexity of Economic System

4.2.1 Diverse Rule Choice of Economic System

An example of diverse rule choice may be cited from *thermodynamical* view. In the earlier era of neoclassical economics, however, they had the idea of an *ex ante* production function as well as *ex post*. Here we can take the idea of *complexity* as *a rule*, indeed. In other words, we introduce complexity as *a new coordinate*. This process will lead us to define a new approach to Evolutionary Economics. We start from the point.[4]

[4]This section is a concise version of Aruka and Mimkes (2005).

Now we can define *the entropy of mixing types* as follows:
N different elements: $N_1, N_2, \cdots, N_n$.
M different classes (or types): $M_1, M_2, \cdots, M_m$.
Entropy S then is defined by the use of the probability P of the distribution of the N elements in M classes of categories: $S = N \ln N - N_i \ln N_i$

In the cyclic process of economic production $\{A \to B \to A\}$, the first part $A \to B$ indicates *productive arrangements* while the second part $B \to A$ does the process of *expenditure*

$$-\oint \Delta W = \oint \delta q = \int_A^B \delta q_{(1)} + \int_B^A \delta q_{(2)} = \int_A^B \delta q_{(1)} - \int_A^B \delta q_{(2)} = \Delta q$$

In *the first law of thermodynamics*, economic net output Δq is only possible by work or production. But it is impossible to calculate *ex ante* how much work we will gain, because the output depends on the production process and there are so many possibilities for this process. *Ex post* we may calculate the net output Δq, as we then know which process has been carried out. *The second law* tells us about the integrating factor T (temperature) that will transform a not exact differential form into an exact form $dS = \delta q/T$. This means the function S is independent of the path and may be calculated *ex ante*.[5]

Under these macroscopic laws, we have different cycle paths, as shown in Fig. 4.2. We shall mainly investigate the way how a path could appear by means

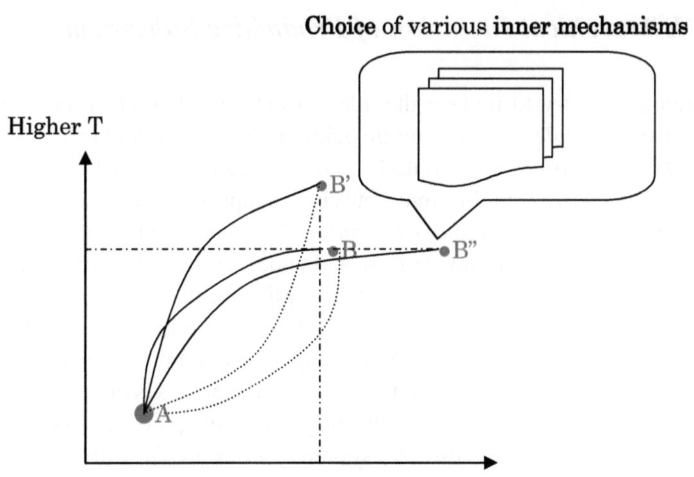

Fig. 4.2 Different cycle paths with different inner mechanisms

[5] S represents the different possibilities for production activities (processes). And we later find $S = \ln P$. In a finite system of N elements the number of possibilities may be calculated by the law of combinations $P = N!/N$.

of inspecting each *inner mechanism* accompanying each different path. That is to say, we focus on the link between the selection of productive activities and the integration of subsystems (*hierarchical inclusion* as defined later).

4.2.2 Truncation and Macroscopic Orders

In the event, we will then have the insight that:

(a) A way to take a Carnot path of production may depend on the size of Δq as well as T. It must be important for Δq which coordinate we could cut the cycle $\{A \rightarrow B \rightarrow A\}$, at which value of B we could choose.
(b) *The truncation of production system* implies a kind of admissible number of combinations of production activities (processes). The feasibility of *truncation* decisively depends on the rate of profit (the rate of surplus) and a price system chosen under a given rate of profit. The precise analysis was smartly given by Schefold (1989). As we see soon in Aruka and Mimkes (2005), the truncation of production system in association with the idea of *hierarchical integration* can lead to *complexity of production*.
(c) Hence we have $(a) \Leftrightarrow (b)$.

4.2.3 Hierarchical Inclusion of Productive Subsystems

In our attempt, we try to replace this idea of entropy of production with a more concrete idea of *complexity* in economic point of view. In particular, we use the idea of *hierarchical inclusion*. This could be summarized in Table 4.1.

We can characterize four different states by a couple of factors, i.e., low and high, on the coordinates of *entropy* and *temperature*. We find there both a state associated with a high temperature but low entropy and another state with a high temperature. This kind of characterization may easily be applied on the plane of *complexity* and *average welfare* as shown later in Fig. 4.3. Thus we notice the case of a simpler production with higher profitability. Hence higher profitability does not necessarily require a more complex production. Observe an economic system of a higher GDP combined with a simpler order as indicating a country producing only crude oil. There may, on the other hand, also be a system of a lower GDP with a more complex system, which implies a system of production by means of *many* commodities. In a more integrated view of production, thus, we can regard each system as a *subsystem*. In this case, we may have an *ensemble* of economically *revealed* subsystem.

If we are facing an optimization problem like maximization of the net domestic income subject to welfare constraints, we could often encounter a set of solution of single process operation like *dictatorship* as pointed out by Aruka (1996). If we should find one of the most efficient activities and invest all the wealth into it,

Table 4.1 Complexity and profitability

Average welfare	Complexity	
	Simple	Complex
Lower	$\{\alpha_1\}^1$	$\{\alpha_1, \alpha_2\}^2$
Higher		$\{\alpha_1\}^1 \cup \{\alpha_1, \alpha_2\}^2$ *Hierarchical inclusion*

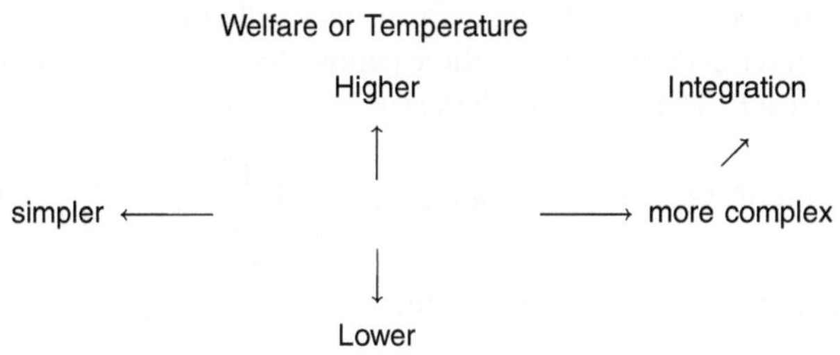

Fig. 4.3 Introduction of complexity and temperature as the new coordinates

we can realize the maximal net income. This answer intuitively seems trivial. Hence, we may imagine that a much simpler system s with higher income often appears or be more probable, compared with a more complex system c:

$$Pr(\omega^{higher} | s) > Pr(\omega^{higher} | c).$$

On the other hand, as for complexity, it holds

$$Pr(\omega^{lower} | s) < Pr(\omega^{lower} | c).$$

On average, it then holds

$$<\omega^s> > <\omega^c>.$$

It is noted that a more complex production could lead to an increase of *the number of interaction* of nodes or productive activities, if we introduced a *recyclic* production of economic system. In what follows, we adopt the next hypothesis on *hierarchical inclusion*: A simpler subsystem with higher profitability could be integrated by means of complementation of more complex subsystem.[6]

[6] Aruka and Mimkes (2005), by the use of von Neumann-Sraffa model of production (von Neumann 1937, Sraffa 1960), attained to the next result:

Theorem 1. *The probability of multiple truncations compatible with a given rate of interest r must be augmented if the number of truncations increases in the range of $g \leq r$. Average welfare could then be risen by a hierarchical inclusion of a single process operation of higher profitability, i.e., by an increase of complexity.*

4.3 Different Expectations by Heterogeneous Agents and the Effects of Neighbouring Agents

4.3.1 A Traffic Problem by Type Selection

According to Strang (1991, pp. 211–213), we have the kind of confusion about "expected" group size. Now suppose, for instance, that there were 10 cars given, with a single driver in three cars, and three pepole, including a driver, in the remaining seven cars each. The expected class size *in view of visitor* is:

$$\begin{aligned}&\text{the summation of 3 cars with a single person}\left\{1\left(\frac{1}{24}\right)+1\left(\frac{1}{24}\right)+1\left(\frac{1}{24}\right)\right\}\\&+\text{the summation of 7 cars with three persons}\left\{3\left(\frac{3}{24}\right)+\cdots+3\left(\frac{3}{24}\right)\right\}\\&=2.75 \text{ persons in the car.}\end{aligned}$$

This average is just the average number of persons per a car that a *random* visitor can expect. On the other hand, the average number of persons which a *random* city authority or a policy maker can expect is

$$\frac{1\times 3+3\times 7\ \text{(persons)}}{10\ \text{(cars)}}=2.4 \text{ per car.}$$

4.3.2 An Expectation Failure

A random visitor feels to be in a more crowded car than in the average car which the city authority estimates. In this setting, as (Strang 1991, pp. 211–213) also provided us, it holds the next relation:

> Traffic problems could be determined by raising the average number of people per car 2.5, or even 2. But that is virtually impossible. Part of the problem is the difference between (a) the percentage of cars with one person and (b) the percentage of people alone in a car. Percentage (b) is smaller. In practice, most people would be in crowded cars.

$$\text{(a)}\frac{3}{10}\gg\text{(b)}\frac{3}{24}.$$[7]

That is to say, the difference between (a) and (b) gives a random visitor a dominant motivation to switch from a fellow passenger to a driver.

[7] We can easily prove the next proposition: With groups of sizes $x_1, x_2, \cdots, x_n$ adding to $G=\sum_i^n x_i$, the average size is $\frac{x_i}{G}$. The chance of an individual belonging to group 1 *is* $\frac{x_1}{G}$. The expected size of the group is $E(x)=x_1\frac{x_1}{G}+\cdots+x_n\frac{x_n}{G}$. Hence it then holds that $\sum_{i=1}^{n}\frac{x_i^2}{G}\geq\frac{G}{n}$.

4.3.3 *Effects of Neighboring Behavior*

The above instance in *type selection* may all give a heuristic finding of *distortion* from the weighted mean of total size. We can enumerate another similar situation as we like. In a random state, whether in a traffic affair, or in a Prisoners' Dilemma game situation, every random participant may decisively depend on *the neighboring type sizes* to select her own choice of transitions, as long as the change of types cannot hurt his or her individualistic gain. The impulse for adaptation to another type could always be generated in our situation.

The motivation due to this kind of distortion to move may be observed by connecting with a kind of *passions within reason.*[8] Transition may be justified in view of his or her passion. Put another way, *a generation of transition* may have an *individualistically rational* ground.

4.3.4 *The View of the Nash Bargaining Problem*

We notice that our types cannot be "divisible". This world constitutes just the field of the Nash bargaining problem indicated by Nash (1951). The types of car situation are 24 kinds in the above traffic problem, although the seating capacity of sedan is normally limited. In fact, some truncation will be needed for a practical use. Each way of which type a random driver can take may be taken as a lottery (Table 4.2).

Thus the total number of the lotteries is

$$\sum_{r=1}^{24} \frac{n!}{r!(n-r)!} = 16,777,215 \text{ ways.}$$

This gives us just the idea of distribution of n balls in r boxes $1, \cdots, n$ as leading to the idea of statistical mechanics. The first part (a) of Fig. 4.4 implies that there are three different coloured balls, three different sized boxes, while the number of the light gray ball is 4, the dark gray's is 5, the black's is 4. In Fig. 4.4b, we have two types of gender (*colour*) and three different subgroups (*box*), but the number of successful couple (*marriage*) on each gender never rises if it never permits *transition* of the members into another box.

Table 4.2 Size combinations

Car	Size 1	Size 2	...	Size 23	Size 24
	${}_{24}C_1$	${}_{24}C_2$	...	${}_{24}C_{23}$	${}_{24}C_{24}$[a]

[a] ${}_nC_r = \frac{n!}{r!(n-r)!}$ is the binomial coefficient. ${}_{24}C_{24}$ is empty for a random driver, because there is not any single driver

[8] Robert Frank loves to discuss plentiful discussions on this kind. See Frank (1988).

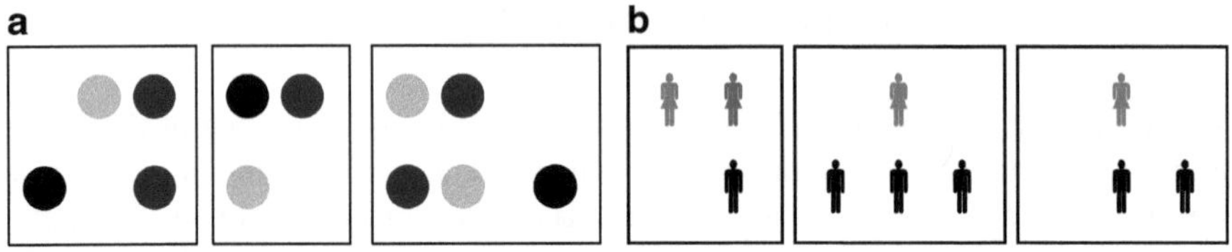

Fig. 4.4 Distribution of r balls in n boxes

4.3.5 *Our New Focus on the Interaction Among Heterogeneous Agents with Many Finite Numbers*

In the Nash bargaining problem, the players are like two boys only allocating "indivisible stationeries." They are not random players. Many agents never appear. In the case of featuring "Nash bargaining solution," Nash thus assumes that the players' preference over lotteries obey the von Neumann–Morgenstern axioms and hence can be represented up to positive affine transformations by a pair of von Neumann–Morgenstern utility functions.[9] We therefore anticipate that here is a certain branching point to go to a new theory of interaction. It must be noticed that Nash pertinently formulates the interaction among *just two boys*. Another way may be given from the view of interaction among *heterogeneous boys* with many finite numbers.

Give 16,777,215 ways, *a huge number* of possibility, we may rather regard these respective events equally happening. One idea is to assume that all the possibility could equally occur. This idea will lead to introducing an idea of *Gibbs distribution*, for instance. We thus rather focus a distribution, i.e., a macroscopic attribute.[10]

4.3.6 *The Idea of Statistical Mechanics and the Boltzman Distribution*

We can now illustrate the Boltzman distribution in view of statistical mechanics. We furthermore assume not only that here are a set of N_k balls in the boxes k, but also that the boxes are endowed with volumes V_k and values E_k. We then have Fig. 4.5, such as a distribution of N balls in K boxes of different volume V_k and value E_k. Here values E_k may be interpreted with energy or price.

---

[9]Recently, "pessimistic subjectivity" is taken account of in the theory of expected utility theory. The theory of this type is called *Choquet* expected utility theory. But we rather are interested in the interaction of heterogeneous agents. See Bassett et al. (2004), for example.

[10]Finally, we must notice *another point*. Even in a society where there are only 24 persons, we always are faced with too many menu lists on type-size allocation when we must select. Suppose we always given an initial allocation of types. We can then always hold the initial allocation as a *status quo* (standstill) policy. In the case of Nash bargaining, the allocation is called the *threat* or *impasse point*.

Fig. 4.5 Distribution of N balls in K boxes of different volume V_k and value E_k

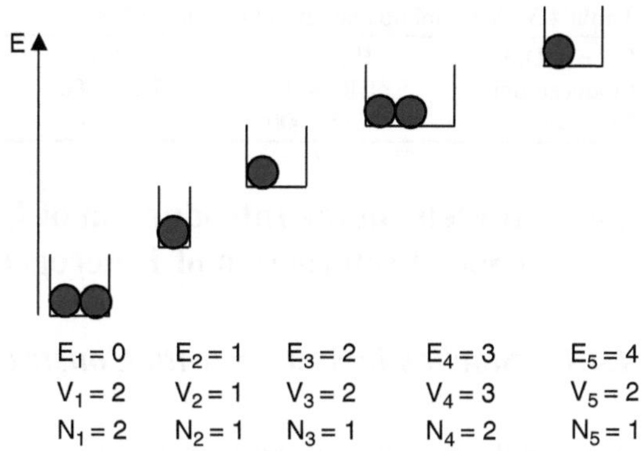

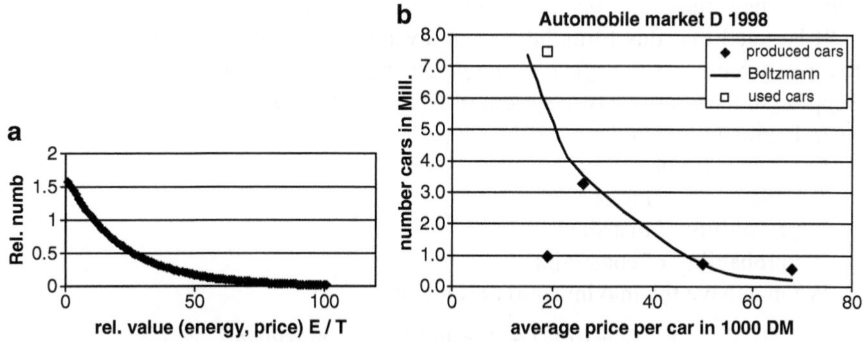

Fig. 4.6 Boltzman distribution

We denote $\sum_k V_k$ by V, and $\sum_i N_i$ by N. We then construct the following Lagrange maximization problem with constraints:

$$L(N_i) = E_0 + N_i E_i - T\{N \ln N - N_i(\ln N_i - \ln(V_i/V)\} \Rightarrow \max!$$

Solving this problem,[11] it follows *the Boltzman equilibrium distribution*:

$$\frac{N}{N_i} = \frac{V_i}{V} \exp \frac{-E_i}{T}$$

This distribution exhibits the relationship between the relative value and the relative number. We produce *the Boltzman distribution* of the automobile market in Germany 1998. Cars are graded by the level of price. We employed the figures of four different classes of sales price measured in DM:19,000, 26,000, 50,000, 68,000 (Fig. 4.6; Table 4.3).

[11] $\frac{\partial L}{\partial N_i} = 0$ implies $E_i - T\left(\ln \frac{N}{N_i} - \ln \frac{V_i}{V}\right) = 0$. Hence $\ln \frac{N}{N_i} \frac{V_i}{V} = -\frac{E_i}{T}$.

Table 4.3 Automobile market in Germany 1998

Price in DM	19,000	26,000	50,000	68,000
Produced units	940,000	3,250,000	720,000	540,000
Used cars	7,450,000			

4.4 An Elementary Introduction of Evolutionary Theory to Social Interaction of Heterogeneous Agents

4.4.1 Stochastic Systems with Constrains

We have defined *entropy* and *temperature* used it to an economic system as *complexity* and *welfare*. Now we apply the constraint to entropy, according to Lagrange, and can then study the stochastic systems with constraints.

It is noted that this formulation can always be activated in *cellular automata* to directly represent *the effects of neighbouring agents*, as Mimkes (2003) showed. Suppose to have a stochastic system of N_i *heterogeneous* elements:

L: Lagrange function.

$E(N_i)$: constraint.

T: Lagrange parameter.

P: probability of distribution.

P_E: probability of constraint.

We then have the maximization problem:

$$L = E(N_i) + T \times \ln P(N_i) \Rightarrow \textit{maximum}!$$

Here we introduce the relative size of group i:

$$x_i = N_i/N.$$

We then have the next relationships:

$$\begin{aligned} \ln P &= N(x_i \ln x_i). \\ E &= N x_i \varepsilon_i + \varepsilon_{ik} x_i x_k + \cdots \end{aligned} \tag{4.1}$$

Expanding $E(x_i N)$ into Taylor series, we obtain the last equation (4.1). We thus have the following maximization problem:

$$L = N\{x_i \varepsilon_i + \varepsilon_{ik} x_i x_k - T(x_i \ln x_i)\} \rightarrow \textit{maximum}!$$

4.4.2 The Simplest Case: Interaction Between the Binary Agents

We illustrate the simplest case where there are only two heterogeneous agents called A and B. N_A is the number of A, and N_B is the number of B. In this case, we can put the efficient as follows:

$$E = N_A p_A E_{AA} + N_A p_B E_{AB} + N_B p_A E_{BA} + N_B p_B E_{BB},$$
$$p_B = N_B/N = x,$$
$$p_A = N_A/N = (1-x).$$

Rearranging the above relations,

$$\ln P = N\{-x\ln x - (1-x)\ln(1-x)\},$$
$$E(x) = N\{E_{AA} + x(E_{BB} - E_{AA}) + (E_{AB} + E_{BA}) - (E_{AA} + E_{BB})x(1-x) + \cdots\}.$$

We put

$$\varepsilon = (E_{AB} + E_{BA}) - (E_{AA} + E_{BB}).$$

We then have the maximization problem in the binary interaction:

$$L = N[E_{AA} + x(E_{BB} - E_{AA}) + \varepsilon x(1-x) - T\{x\ln x + (1-x)\ln(1-x)\}] \Rightarrow \max!$$

Now, by the use of ε, we define the characteristics of binary interaction as follows:

Cooperation: $\varepsilon > 0$.
Integration: $\varepsilon = 0$.
Segregation: $\varepsilon < 0$.

4.4.3 Six Real Structures in Binary Agent Systems

We can rearrange the above problem as follows:

$$L(x,T) = L_0 + \varepsilon x(1-x) - T\{x\ln x + (1-x)\ln(1-x)\} \Rightarrow \textit{maximum}!$$

Here $\varepsilon = (E_{AB} + E_{BA}) - (E_{AA} + E_{BB})$ is the *structure parameter* of phases. The problem can easily be solved into:

$$\frac{T(x)}{\varepsilon} = \frac{(1-2x)}{\ln x - \ln(1-x)} \tag{4.2}$$

This solution can depict *the phase diagram* for binary alloys (e.g., *Au-Pt*). According to the conditions of coefficient ε and E_{ij}, we now summarize such six real structures in binary agent systems as produced in Table 4.4.

4.4.4 An Example: Intermarriage Interaction

A binary relation can have the percentage dynamics

$$P(x) = 2x(1-x). \tag{4.3}$$

Table 4.4 Six real structures in binary agent systems

Structure parameter	Society	Nature
$\varepsilon<0$ and $E_{AB}+E_{BA}>0$	Segregation	Alloy, real liquid
$\varepsilon<0$ and $E_{AB}+E_{BA}<0$	Aggression	Chemical reaction
$\varepsilon>0$ and $E_{AB}=E_{BA}>0$	Partnership	Compound
$\varepsilon>0$ and $E_{AB}\neq E_{BA}>0$	Hierarchy	Solid
$\varepsilon=0$ and $E_{AB}+E_{BA}>0$	Democracy	Ideal liquid
$\varepsilon=0$ and $E_{AB}+E_{BA}=0$	Global structure	Gas

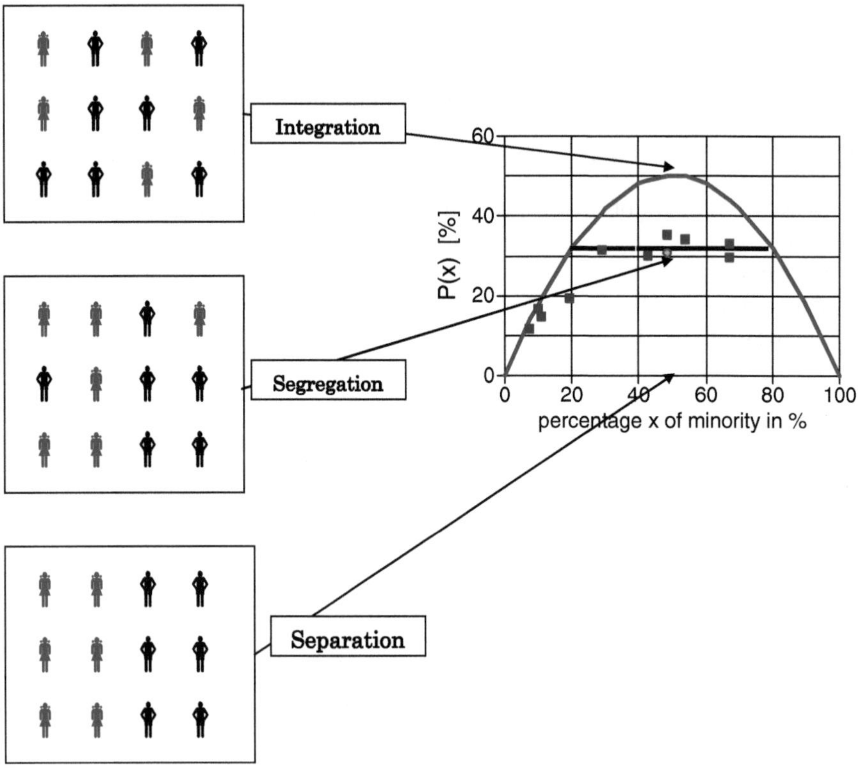

Fig. 4.7 Intermarriage as a thermometer of social temperature. • denotes the mark of Switzerland (*CH*: Confederation Helvetica) which is located approximately on the *centre* of the real line

Let x be the proportion of *minority*, for instance. If x approaches 0.5, its percentage must be 0.5; If it exceeds, it cannot by itself be a minority. We take an example of *intermarriage* in Europe, in particular, in Germany. If $\frac{T}{|\varepsilon|}>1$, *ideal integration* then dominates; If $\frac{T}{|\varepsilon|}<1$, the segregation then dominates; If $\frac{T}{|\varepsilon|}\rightarrow 1$, *separation* then dominates; *Social Temperature T* is given by (4.2). Difference of temperature characterizes social state. If $\frac{T}{|\varepsilon|}$ is quite low, the social state must be aggressive since the separation of population dominates (Fig. 4.7).

References

Anderson P, Arrow K, Pines D (eds) (1988) The economy as an evolving complex system: studies in sciences of complexity. Addison-Wesley, Redwood, CA

Aruka Y (1996) Dictatorship, price hike like bubble and employment adjustment on the non-goldenpath of the von Neumann growth model with joint-production. In: Proceedings of econometric society Australasian meeting, vol 1. University of Western Australia, Perth, pp 475–504

Aruka Y, Mimkes J (2005) Complexity and interaction of productive sub-systems under thermodynamical viewpoints, systems under thermodynamical viewpoints. In: WEHIA2004: the 10th workshop on economics with heterogeneous interacting agents, University of Essex, UK, 13–15 June 2005

Bassett G, Koenker R, Kordas G (2004) Pessimistic portofolio allocation and choquet expected utility. CEMMAP Working Paper CWP09/04

Bellman R (1961) Adaptive control processes: a guided tour. Princeton University Press, Princeton, NJ

Bierlaire M (1997) Discrete choice models, mimeo

Frank RF (1988) Passions within reasons. The strategic roll of the emotions. W.W. Norton, New York (Paper edition 1989; German translation, Orednbourg 1992; Japanese translation, Saiensu-sha 1995; Greek translation, Kastoniosis 2000)

Hodgson GM (2003) The preface to Japanese version of "Economics and evolution: bringing life back into economics". Polity, Cambridge

Holland JH (1992) Adaptation in natural and artificial systems. MIT Press, Cambridge, MA

Holland JH (1995) Hidden order: how adaptation builds complexity. Perseus Books, Reading, MA

Hoover KD (2001) The Methodology of empirical macroeconomics. Cambridge University Press, Cambridge, UK

Keynse JM (1972) The end of laissez-faire, The collected writings of John Maynard Keynse. IX Essays in Persuation. Macmillan, London

Luce R (1959) Individual choice behavior: a theoretical analysis. Wiley, New York

Mimkes J (2003) Thermodynamics of economic agent systems, Part III. University of Paderborn, Paderborn (mimeo)

Nash JF (1951) Noncooperative games. Ann Math 54:286–95

Schefold B (1989) Mr. Sraffa on joint production and other essays. Unwin Hyman, London

Sraffa P (1960) Production of commodities by means of commodities. Cambridge University Press, Cambridge, UK

Strang G (1991) Calculs. Wellesley-Cambridge Press, MA

von Neumann J (1937) Über ein Ökonomisches Gleichungssystem und eine Verallgemeinung des Brouwerschen Fixpunktsatzes, Ergebnisse eines Mathematischen Kolloquiums 8:73–83 (translated as "A model of general economic equilibirum." Rev Econ Stud 13:1945–1946)

Weidlich W (2000) Sociodynamics: a systematic approach to mathematical modeling in the social sciences. Harwood Academic, London

Chapter 5
The Law of Consumer Demand in Japan: A Macroscopic Microeconomic View

Yuji Aruka

Abstract In economics, there have been found few fundamental theories as complying with statistical laws, really examined by observation. Econometrics still clings to a very special structure, patronized with favoritism by the traditional economic theory. In 1990s, Hidenbrand, Grandmont, Grodal and others have tried to formulate consumer demand by a procedure definitely different from the traditional way of using an individualistic utility function. An alternative approach is to make assumptions on the population of the households as a whole – *the macroscopic microeconomic approach*. If the households' demand functions are not identical then one needs a certain form of *heterogeneity* of the population of households. This approach really also requires the empirical tests. The method of nonparametric test on income distribution to estimate covariances on the households spending may be applied. In this article, Japanese Family Expenditure Data is used for estimation. The law of consumer demand is one of the most important topics since a birth of economics. Our *statistical economics* should, first of all, demand innovation in this area.

5.1 The Law of Consumer Demand

5.1.1 Slutsky Equation

If we should derive the law of consumer demand in terms of the individual demand functions (or individual preferences), we could be forced to employ some restrictive assumptions such as that individual demand functions are linear in income, i.e.,

$$f^i(p,x^i) = x^i f^i(p,1).$$

Y. Aruka
Faculty of Commerce, Chuo University, Higashinakano, Hachioji, Tokyo 192–0393, Japan
e-mail: aruka@tamacc.chuo-u.ac.jp

Y. Aruka (ed.), *Complexities of Production and Interacting Human Behaviour*,
DOI 10.1007/978-3-7908-2618-0_5, © Springer-Verlag Berlin Heidelberg 2011

Otherwise, we contented ourselves with an *ad hoc* assumption of gross substitutability to be imposed on the Slutsky equation. The individual demand function is usually expressed as the form of Slutsky equation:

$$\frac{\partial f_j^i(p,x^i)}{\partial p_k} = \frac{\partial h_j^i(p,x^i)}{\partial p_k} - \frac{\partial f_j^i(p,x^i)}{\partial x^i} f_k^i(p,x^i) \tag{5.1}$$

Total Demand Effects = Substitution Effects *plus* Income Effects.

A compensated income h[1] is defined as the original demand $f(p, x(p))$ after a footnote size prices variation from p to q

$$x(q) = x(p) - (p-q)f(p,x(p)). \tag{5.2}$$

The assumption of gross substitutability implies that a footnote size price variation is always held positive.

Without an *ad hoc* hypothesis such as gross substitutability, Hildenbrand has been successful to introduce the form of *heterogeneity* of the population of households by constructing a certain set of macroscopic hypotheses. We can only give a brief summary on them here. We denote the population of households by I. The households can be classified into the groups of different income x. The households with income x_i are called the x_i-household. The x_i-household may also be called the x_i-cloud, because each member in the x_i-household has a various vector of demand $f^i(p,x^i)_{i\in I}$ where $f^i(p,x^i)$ is demand of each household i. The market demand $F(p)$ thus is an average idea of household demand, the sum of household demands divided by the total number of households. That is,

$$F(p) = \frac{1}{\#I}\sum_{i\in I}\left[f_j^i(p,x^i)\right], \tag{5.3a}$$

$$S(p) = \frac{1}{\#I}\sum_{i\in I}\left[\frac{\partial h_j^i(p,x^i)}{\partial p_k}\right], \tag{5.3b}$$

$$A(p) = \frac{1}{\#I}\sum_{i\in I}\left[\frac{\partial f_j^i(p,x^i)}{\partial x^i} f_k^i(p,x^i)\right]. \tag{5.3c}$$

A Jacobian matrix of the market demand function $F(p)$, considering the so-called Slutsky decomposition, can be written in the form of

---

[1]In view of an individualistic utility function, h can be derived from the problem.

$$h^i(p,x^i) := \arg\min_{u(z)=\xi} pz,$$
$$h(q,x(q)) : q \to f(q,x(q)).$$

$$\frac{\partial F(p)}{\partial p} = S(p) - A(p), \tag{5.4}$$

by which it must hold the condition

$$\frac{\partial F(p)}{\partial p} < 0 \tag{5.5}$$

in order to assure the *Law of Demand* on average.

5.1.2 An Alternative Derivation of the Law of Demand by Hildenbrand

The following hypotheses supporting the above condition are smartly proposed by Hildenbrand (1994, pp. 19–21).

Hypothesis 1. The average Slutsky substitution effect matrix *S(p)* is negative semidefinite.

Hildenbrand is not interested in empirical testing on the Hypothesis 1. He instead discussed the assumptions being compatible with *generic function* to be defined. See Hildenbrand (1994, Chap. 2).

Hypothesis 2. The average income effect matrix A is positive semidefinite.

We can test *Hypothesis 2* empirically by calculating the matrix of second moments of each clouds of households demand underlying *the Family Expenditure Data* officially published. The second moments can be utilized an index of *spread*. In other words, *spread* may be interpreted a measure of the heterogeneity of households. The cloud $\{y^i\}_{i\in I}$ is judged *more spread*, if $m^2\{y^i\}_{i\in I} > m^2\{z^i\}_{i\in I}$.

Hildenbrand then proved quite ingeniously that *Hypothesis 2* can be equivalent to *Hypothesis 3*, provided the demand functions satisfy the budget identity. See Hildenbrand (1994, Appendix 5).

Hypothesis 3. [Increasing spread and expanding dispersion of household demand] For every sufficiently small $\Delta > 0$, *the cloud* $\{f^i(p, x^i + \Delta)\}_{i\in I}$ is *more spread* than $\{f^i(p, x^i)\}_{i\in I}$

Dispersion on the clouds can be defined to measure the heterogeneity of households. To remove the so-called Giffen effects, the average income effect matrix A must be positive semi-definite. According to a new theory of Hildenbrand, this corresponds to the fact that "dispersion in each income to measure the heterogenity of households may increase, as the income size increase". In short, it then holds:

$$\mathrm{cov}\{x^i + \Delta\}_{i\in I} - \mathrm{cov}\{x_i\}_{i\in I} > 0, \tag{5.6a}$$

$$\mathrm{Ell}(\mathrm{cov}\ v(x^i + \Delta)) \supset \mathrm{Ell}(\mathrm{cov}\ v(x^i)). \tag{5.6b}$$

Our empirical test thus comes down to calculate the *matrix of covariance* around the mean on the clouds.[2]

5.1.3 An Illustration in the One-Commodity Economy

The one commodity example is a short way for us to understand a new perspective of Hildenbrand's theory. Lewbel (1994, Appendix) gives a compact view on it in terms of the one commodity world. The *Slutzky Equation* in one commodity case is the slope of an individualistic demand curve:

$$\frac{\partial f}{\partial p} = \frac{\partial f}{\partial p} + \frac{\partial f}{\partial x} f - \frac{\partial f}{\partial x} f = s - \left(\frac{1}{2}\frac{\partial f^2}{\partial x}\right)' \tag{5.7}$$

Applying operator E to averaging among the individual agents,

$$\frac{\partial E(f)}{\partial p} = E(s) - \frac{1}{2} E\left(\frac{\partial f^2}{\partial x}\right) = E(s) - \frac{1}{2} E[R'(x)] \tag{5.8}$$

(1) *Hypothesis* 1 implies $E(s)<0$. (2) $R(x)$ is $E(f^2)$, the square mean of various demands of the income level x, since $R(x) = E(f^2)$.[3]

$$R(x) = \underbrace{(E(f^2) - \{E(f)\}^2)}_{\text{variance}} + \underbrace{\{E(f)\}^2}_{\text{square mean of } f} = E(f^2). \tag{5.9}$$

The Larger the value of $R(x)$, the more the diversity or heterogeneity of household demands at the similar income levels will be. We can easily know the effects of variance in this model. $R(x)$ therefore measures the *spread* of household demands. Given the density function of the income distribution ρ, in general,

$$E[R'(x)] = \int R'(x)\rho(x)dx = \int E\left(\frac{\partial f^2}{\partial x}|x\right)\rho(x)dx. \tag{5.10}$$

$E[R'(x)]$ depends on the rank of demand and also on the configuration of income distribution. These features are really demanding the empirical analysis.

---

[2]Ell means the ellipsoid of dispersion in (5.19) of Sect. 5.2.5.

[3]$R'(x) = \frac{\partial R(x)}{\partial x} = \frac{\partial E(f^2)}{\partial x}$. Then $R(x) = E(f^2)$.

5.2 Statistical Tests

5.2.1 The Estimation of the Income Distribution in Japan

We are faced with a limited source of data in income distribution. In Japan, we can only obtain the data of at most 18 classes *ranked* in family income. Officially, there is nothing more precise than this. Even in this case, we must collect these from each annual report since 1979 to make out the annual sequential data. Furthermore, it is to be noted that there is a discontinuity in the classification size of 18 classes because the income bandwidth employed has changed since 1991. In fact, the income size among individuals has increased much more in the late 1980s due to the final stage of the so-called bubble economy. Thus it must be noted that we are forced to employ another classification in size of the income data for the periods since 1991.

5.2.2 The Covariance Matrix under the Average Derivative Method

This task is closely connected with the empirical testing of *Hypothesis 3*. In fact, Hildenbrand, in cooperation with Haerdle, and Kneip working in the field of nonparametric testing, estimated the covariance matrices derived from the spending structures of individual households which are available in *UK Family Expenditure Survey (FES), French Enquâte Budget de Famille (EBF)* and the surveys in other developed countries. The essence of this estimation has been the use of *the average derivative method* of Haerdle and Stoker (1989). In order to use this method, he has used the *Rosenblatt–Parzen kernel density estimator*:

$$\frac{15}{16}(1-u^2)^2 \quad \textit{for} \quad |u| \leq 1. \tag{5.11}$$

More precisely for our practical purpose, this can be specified by the use of *bandwidth h* in the following manner:

$$\frac{1}{h}\frac{15}{16}\left(1-\left(\frac{x-X}{h}\right)^2\right)^2. \tag{5.12}$$

Here x denotes income, X the average income. Incidentally, it is noted that the quadratic form after differentiation can leave a cubic form.

Our data source comes from the volumes of the *Annual Report on the Family Income and Expenditure* issued by the Statistics Bureau, Management and Coordination Agency, Government of Japan. In each volume, we can find the pages of Table 4 in the recent editions of the *Yearly Average of Monthly Receipts and*

Table 5.1 Income clouds in 1998

Class[a]	1	2	3	4	5	6	7	8	9	10	11	12	13	14	15	16	17	18
Income[b]	153	226	275	324	373	422	473	523	573	622	674	722	772	843	946	1,110	1,358	2,011
Household	162	198	322	448	493	491	504	506	483	477	441	402	349	633	477	780	380	395

[a]The class number i implies the i-th ranked income class from the bottom
[b]Income is measured in 10^4

Disbursements per Household by Yearly Income Group (All Household) of the 18 income rank classifications. Table 5.1 is a set of each average income in each income class and its number of the households in the same cloud in 1998.

It is noted that *income* is the average income $X[i]$ in each income class I for $i \in I$, while *household* is the total number $\sum q_I^i$ in each income class I also for $i \in I$. We have replicated a figure of x by the number of $\sum q_I^i$ as if there really existed the average households. q is the total number of households in the concerned year.

$$q = \sum_I \sum_i q_I^i \tag{5.13}$$

5.2.3 The Cross Validation

We select h, the bandwidth of the kernel function to minimize $W(i, j)$, according to Haerdle (1990):

$f(x)$: the sum of the *kernel function* $K(x)$ all over the samples –in our case, a so-called samples, namely replicating each average by the number of households in each income group – when income x is taken a ratio to the total income of the sample. This is to be discounted by $q - 1$, where q is the total households.

$W(i, j)$: the (i, j) element of $K(x)/f(x)$.

Here i shows each item of the spending object, j shows each income size.

In many simulations of

$$minimizing\ W(i,j)$$

on the sample data of Japan for the periods of 1979 to 1998, it seems us better to choose $h = 1.5$.[4]

Then the kernel density will be, if X is 1,000th household's income in terms of the total average income,

$$k(x, X[1000]) = 0.000079(1 - 0.444444(-0.447446 + x^2))^2.$$

[4]As $h = 1.5$ is chosen, the eigenvalues of the covariance matrices remarkably holds all positive during almost the periods of 1979 to1998. See Aruka (2000).

Table 5.2 The spending structure by the representative household in each income class (Asterisks express the values of each income cluster.)

Income	1	2	3	4	5	6	7	8	9	10
Class	Food	Housing	Fuel	Furniture	Clothes	Medical	Transport	Education	Reading	Other
$I_{j\in 1to18}$	*	*	*	*	*	*	*	*	*	*

Thus the kernel density distribution for 1998 will be expressed in the following manner:

$$\rho = \sum_{i}^{q} k(x, X[i]). \tag{5.14}$$

We can then give the derivatives of ρ with respect to x. Hence it follows:

$$L = \frac{D(\rho, x)}{\rho}. \tag{5.15}$$

In our annual report of the family expenditure data, there are the ten consumption aggregates categories of the Living Expenditure $j = 1, \ldots, 10$ as follows: $j = 1$. Food; 2. Housing; 3. Fuel, light, and water charges; 4. Furniture and household utensils; 5. Clothes and footwear; 6. Medical care; 7. Transportation and communication; 8. Education; 9. Reading and recreation; 10. Other living expenditure.

The spending structure Table 5.2 consists of 18 rows multiplied by 10 columns. Underlying this table,

$$v = v(I, j), \tag{5.16}$$

the desired covariance matrix can be generated. By *the average derivative method* of Haerdle and Stoker (1989), we identify the distribution of the x-household income $v(p|x)$ with

$$v \textit{ appreciated by } L = \frac{D(\rho, x)}{\rho},$$

to attain *the covariance matrix* for any smooth income distribution v. That is to say, this may be identical to the

$$C_\rho(p) = \int \partial_x \mathrm{cov}\ v(p|x)\rho(x)dx. \tag{5.17}$$

5.2.4 *The Empirical Tests for Growing Variances*

The two kinds of the estimation results are characterized in terms of the *second moment curves* and *variance curves* respectively. They show the relationship of the diagonal elements of the above matrices on the average total expenditures. In this paper, we only refer to the *variance curves* since the property of *increasing dispersion* is concerned here. Viewing roughly, at the first sight, it seems satisfied that *Hypothesis 3* is empirically viable in the concerned periods of the Japan economy. See Aruka (2000). First of all, we can investigate empirically how the estimates of *the conditional variance curve* behave, as we expected them to be increasing as the expenditures of households increase.

Growing variances has easily been checked without resort to the Rosenblatt–Parzen kernel estimator. I have applied *the normal kernel estimator* to smooth our spending structure table and then calculated the covariant matrix in each year for 1979–1998. See Aruka (2000). The diagonal elements of the covariance matrix, given the correspondence of the variances of the total expenditures, may construct *a variance curve* as Hildenbrand called. It may be easily verified that the *variance* of each commodity aggregate is inclined to *grow* with respect to an increase of size of the total expenditures in the almost intervals under consideration. Due to the restriction of the allocated space, I only show the result of 1998 in Fig. 5.1.

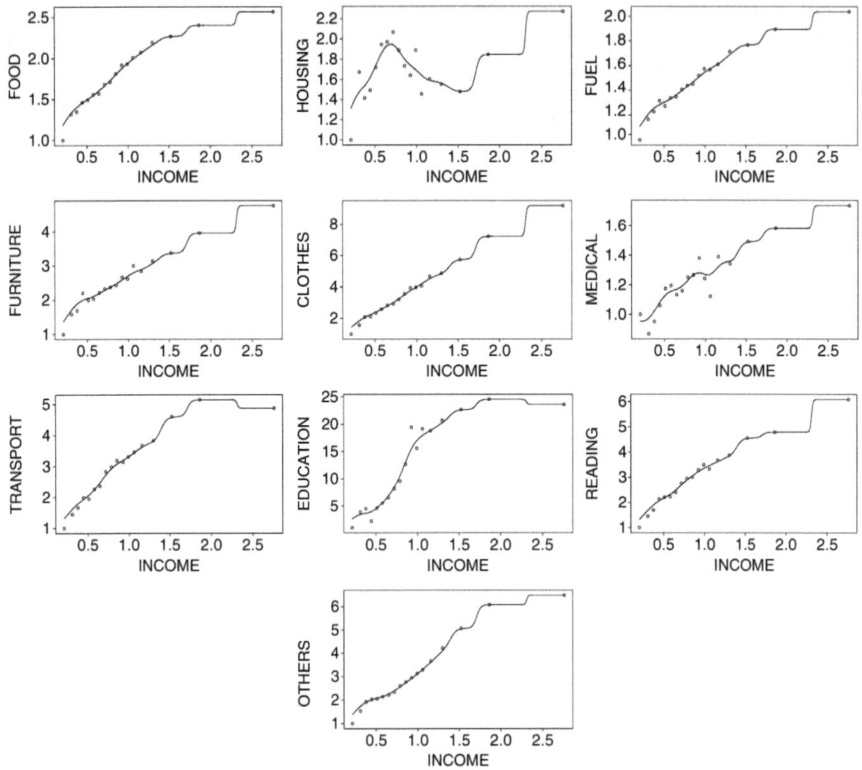

Fig. 5.1 Variance curves in 1998

5.2.5 The Empirical Tests for Expanding Dispersion

Secondly, we may estimate *the ellipsoid of dispersion*. Hildenbrand virtually suggested the use of *average increasing dispersion of x-Households' Demand.*

Property 1: $\int\{\text{cov } v(x+\Delta) - \text{cov } v(p|x)\}\rho dx$ *is positive semidefinite* if $\Delta>0$.

I believe that *Property 1* can be empirically verified because all the eigenvalues are positive in many successive years, although they decisively depend on how to select the bandwidth. The confidence on *Property 1* may be also demonstrated by the following empirical confirmation of *growing dispersion.*[5]

In Hildenbrand (1994, pp. 78–79), we have a concise illustration on *the ellipsoid of dispersion*. Let s_i for $i = 1,2$ be the *standard deviation*, and r be the correlation coefficient which is defined as $r = \dfrac{\text{cov}(1,2)}{s_1 s_2}$.[6] It then holds the covariance matrix

$$\text{cov } v = \begin{bmatrix} s_1^2 & r s_1 s_2 \\ r s_1 s_2 & s_2^2 \end{bmatrix}. \tag{5.18}$$

Thus we have an expression of *ellipsoid* in terms of the variance:

$$q = x \text{ cov}^{-1} \; v \; x = \frac{s_2 x_1^2 - 2 r s_1 s_2 x_1 x_2 + s_1 x_2^2}{(1-r^2) s_1^2 s_2^2}. \tag{5.19}$$

Here we rather used the covariance matrix derived from the average derivative method. Instead of observing all x on the domain of all the households in the economy, let all x be *fixed* at 1, while $\Delta = 0$.[7] It then follows the covariance matrix:

$$\partial_\Delta \text{cov } v(\Delta, p)|_{\Delta=0} =: C(p). \tag{5.20}$$

Now suppose x to be fixed at $1{:}\bar{x} = 1$. We arrange two different cases against $\bar{x} = 1{:}\ x = 0.5\bar{x}; x = 1.5\bar{x}$. It may normally be guessed that the dispersion expands as x increases. We take any two consumption aggregates. I can then show the results of ellipsoids both in 1998 in Fig. 5.2. See Aruka (2000) for another year. In the

---

[5]Since this may be considered $C_\rho = \int \text{cov } v(x,p)\rho dx$ by identifying ∂_x with $\text{cov } v(x+\Delta) -$ $\text{cov } v(p|x)$, *Property 1* virtually is equivalent to say that C_ρ is *positive semidefinite*. Thus the statistical tests of *Property 1* requires to check whether the eigenvalues of C_ρ are *all semipositive* or not. According to Hildenbrand (1994), and Hildenbrand and Kneip(1993), these eigenvalues should be subject to the *bootstrap test*. Our matrix of C_ρ derived by the *average derivative method* has produced all strictly positive eignvalues of 10 distinct roots all in 1990s, if the bandwidth h is specified 1.5. Over a half of the eigenvalues, we may have the eigenvalues of nearly zero. If we should apply the bootstrap test on our covariance matrices C_ρ, there could appear many negative eigenvalues. See Aruka (2000).

[6]The $\text{cov}(1,2)$ denotes (1,2)-th component of the covariance matrix.

[7]This may be regarded as the form C_ρ appreciated at all $x = \bar{x}$.

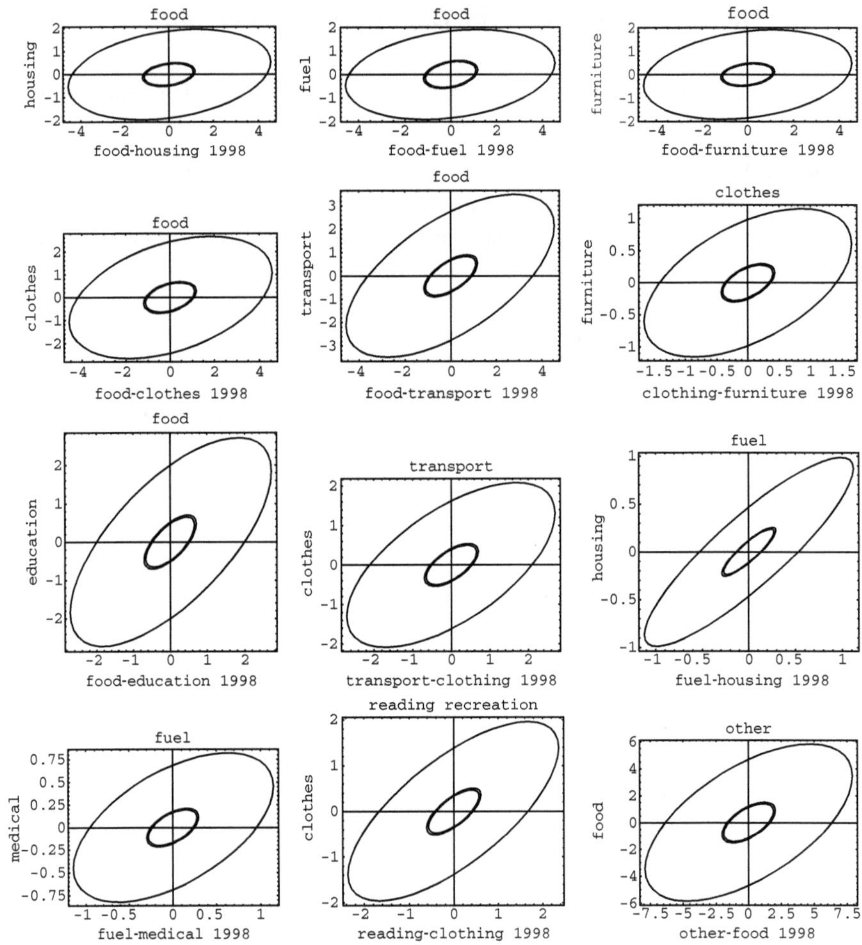

Fig. 5.2 Ellipsoids of dispersion in 1998

concerned framework, three ellipsoids at the levels of $\bar{x} = 0.5, 1, 1.5$ can be depicted. We only notice the fuel–medical ellipsoid in 1998 violating our property in that the inner two circles are crossing.

5.2.6 The Statistical Test of Property 2

Finally, we are ready to verify the other property proposed by Hildenbrand to assure the dispersion of variances:

Property 2: $\partial_\Delta \text{cov}\ v(\Delta, p)|_{\Delta=0}$ is *positive semidefinite* on the hyperplane $F(p)$.

Table 5.3 Eigenvalues in 1998

$\Gamma C \Gamma$	6	0	0	0	0	0	0	0	0	0
C	5.7355	0.329683	0.11656	0.01638	0.00695	0.00356	0.00144	0.00022	0.00009	0.00002

In order to check this property, let Γ be a projection matrix which is orthogonal to C. Applying Gram–Schmidt orthogonirization to C, for instance, we can obtain Γ. It is seen that "the matrix C is positive semidefinite if and only if eigenvalues of $\Gamma C \Gamma$ is positive semidefinite" Hildenbrand (1994, p. 109). Given the covariance matrix C in 1998, then, we can easily calculate the eigenvalues of C (Table 5.3).

Thus *Property 2* has completely been proven in our case.

References

Aruka Y (2000) Family expenditure data in Japan, and the law of demand: macroscopic Microeconomic View, Discussion Paper Series, 9. The Institute of Economic Research, Chuo University, Tokyo

Haerdle W (1990) Applied nonparametric regression. Cambridge University Press, Cambridge, UK

Haerdle W, Stoker T (1989) Investigating smooth multiple regression by the method of average derivatives. J Am Stat Assoc UK 84:986–995

Hildenbrand W (1994) Market demand. Princeton University Press, Princeton

Hildenbrand W, Kneip A (1993) Family expenditure data, heteroscedasticity and the law of demand. Ricerche Economiche 47:137–165

Lewbel A (1994) An examination of Werner Hildenbrand's market demand. J Econ Lit 32:1832–1841

Mitjushcin LG, Polterovich WM (1978) Criteria for monotonicity of demand functions. Ekonomka i Matematicheskie Metody 14:122–128 (in Russian)

Chapter 6
How to Measure Social Interactions Via Group Selection? Cultural Group Selection, Coevolutionary Processes, and Large-Scale Cooperation: A Comment

Yuji Aruka

Abstract It is interesting to know that Henrich's skillful challenge of the Price equation of biology in Price [Nature 227 (1970) 520; Annals of Human Genetics 35 (1972) 485] to locate large-scale cooperation of human beings certainly gives a straight way to argue social interactive mechanism in economic theory. In fact, Glaeser and Scheinkman [E.L. Glaeser, J.A. Scheinkman, Measuring social interactions, in: S.N. Durlauf, H.P. Young (Eds.), Social Dynamics, Brooking Institute, Washington, DC, 2001, pp. 83–131] exhibits a limited form of the Price equation in economic theory without any explicit reference to it. This suggests a common analytical way to measure interactive effects of between-groups and within-group whether in biology or economics. This way however is to be accompanied by some difficulties. Some overarching mechanism between between-groups and within-group will be necessary to establish a definite direction of group selection. The idea of Hildenbrand [W. Hildenbrand, Market Demand, Princeton University Press, Princeton 1994] may be helpful to do this, although his trial has been done in a qualitatively different field.

Keywords Heterogeneities · Macroscopic order · Overarching mechanisms · Social interactions

---

Reprinted from Journal of Economic Behavior and Organization 53(1), Aruka, Y., How to Measure Social Interactions Via Group Selection?: Cultural Group Selection, Coevolutionary Processes, and Large-Scale Cooperation: A Comment, 1–7 (2004), http://www.sciencedirect.com/science/journal/01672681. With kind permission from Elsevier, Oxford, UK.

Y. Aruka
Faculty of Commerce, Chuo University, Higashinakano, Hachioji, Tokyo 192-0393, Japan
e-mail: aruka@tamacc.chuo-u.ac.jp

DOI 10.1007/978-3-7908-2618-0_6,

6.1 Social Interactions in Utility Theory

6.1.1 New Issues in Utility Function

We begin to notice that Henrich's insights of prosocial genes under the cultural transmission mechanism could in general be compatible with a new understanding of evolutionary adaptive economic systems.

A new movement to incorporate social interaction in utility functions of economics has notably begun during these days. Some economists are working to capture the effects of social interaction among the agents. It is not difficult for us to mention a series of new attempts to introduce social interactions and heterogeneities of agents. One set may come about from the random preferences approach given by Hildenbrand (1971), Föllmer (1974), followed by Durlauf (1997, 2000) who borrows the idea of Gibbs distribution in statistical mechanics; while another set is of Weidlich and Haag (1983), and Helbing (1995) who are concerned with application of master equations to social interactions. Aoki (1996) also is to be mentioned.

6.1.2 Cooperative Systems Under a Macroscopic Order

In these models, a macroscopic field on which particular agents interact could bind themselves by the field itself. In other words, some positive feedback (i.e. reinforcement mechanism, passing through a critical point of social situations) may give a limiting system of order, which might be a cooperative system. One type of reinforcement mechanism can come towards bringing cooperation into play via copying or imitating the reference agents (their neighbors), as Arthur (1994) has argued. A source of copying behavior by individuals may come from learning. Voluntary learning by individuals can however not always lead a group to a cooperative situation. Glaeser and Scheinkman (2001, p. 94), similarly as in Henrich, has referred to social interaction through punishment mechanism:

> Social interactions occur because through learning this ideal behavior, individual influences each other. Natural examples of this type of effect occur in crowd behavior where individuals seem to completely forego what is commonly thought of as civilized behavior because they are sanctioned by the crowd (see, e.g. the extensive literature on the motivation of Nazis).

6.1.3 Sub-Grouping and the Memes of Subgroups

Another important feature in social interaction of agents is found in partitioning or labeling individuals into their sub-groups. Group selection through its cultural transmission is relevant to the concept of meme or social norms. In this context, Friedman and Sighn (2001, pp. 106–107) argued vengeance as a negative reciprocity:

> All known groups of humans maintain memes that prescribe appropriate behavior toward fellow group members, and typically prescribe different appropriate behavior toward individuals outside the group. ... The success of the meme, as with any other adaptive unit, is measured by its ability to displace alternatives, by its fitness. ... Groups affect individual fitness in several ways. ... they provide gains from internal cooperation and some gains from external cooperation.

6.2 A Simple Model of Local Interactions

6.2.1 Glaeser and Scheinkman

With the above comprehensive background in line with Henrich's stated circumstances, Glaeser and Scheinkman (2001) gives a simple model of local interactions for their practical purpose to measure social interactions, although in a very restricted model.[1]

People respectively live in different cities. θ is a distribution of quality across people, which has mean zero. Its variance is denoted σ_θ^2. θ_i is the individual i's quality, assumed to be constant across cities. X_i represents individual-level characteristics like age or gender. The idiosyncratic tastes of individual i may then be written:

$$\Theta_i = \theta_i + f(X_i). \tag{6.1}$$

"[T]he marginal utility of the action for individual i is directly by an idiosyncratic taste shock Θ_i, and his neighbor's action A_{i-1}". (Glaeser and Scheinkman 2001, p. 95):

$$U(A_i, A_{i-1}, \Theta_i) = \Theta_i A_i - \frac{1}{2}(1-\alpha)A_i^2 - \frac{1}{2}\alpha(A_i - A_{i-1})^2. \tag{6.2}$$

More specifically, the actions space is supposedly restricted on the real line. Agents may live in their respective cities j on the real line. The choice of action by individual i is based either on his own taste for the action or by imitating his predecessor $i-1$ on the domain.[2]

Individual i chooses an action A_i based on their idiosyncratic tastes Θ_i, and their predecessors' action level A_{i-1}. The individual i's action A_i thus is defined:

$$A_i = \theta_i + f(X_i) + \alpha A_{i-1}. \tag{6.3}$$

---

[1]The next two subsections are a brief summary on Section 3 of Glaeser and Scheinkman (2001, pp. 94–100).

[2]The benefit from copying their predecessor will be measured by $(1-\pi)$.

Let $\overline{A}_j$ be the mean action level in city j. Equivalently[3]:

$$A_i - \overline{A}_j = \theta_i + f(X_i) - \overline{f(X)}_j + \alpha(A_{i-1} - \overline{A}_j). \quad (6.4)$$

Here α is regarded as the regression coefficient of A_i with respect to A_{i-1}. It is also assumed that "there is no sorting across neighborhoods within cities". (Glaeser and Scheinkman 2001, p. 95) Notice that offsprings of i never break out in this line.

6.2.2 *Local Interactions*

We can then calculate the variance of $A_i - \overline{A}_j$. In equilibrium, it holds:

$$\mathrm{Var}(A_i - \overline{A}_j) = \mathrm{Var}(A_{i-1} - \overline{A}_j). \quad (6.5)$$

$\mathrm{Var}_j f(X)$ represents the variance of $f(X)$ within city j.[4] In addition, $\mathrm{Var}_j f(X)$ is assumed to be constant across cities:

$$\mathrm{Var}(A_i - \overline{A}_j) = \sigma_\theta^2 + \underbrace{\mathrm{Var}_j f(X)}_{\text{within-city}} + \underbrace{\alpha^2 \mathrm{Var}(A_{i-1} - \overline{A}_j)}_{\text{between-cities}} = \frac{\sigma_\theta^2 + \mathrm{Var}_j f(X)}{1-\alpha^2}. \quad (6.6)$$

In the limit, Glaeser and Scheinkman (2001) gives the regression coefficient:

$$\alpha = \frac{\mathrm{Var}_{\mathrm{agg}} - \mathrm{Var}_{\mathrm{ind}}}{\mathrm{Var}_{\mathrm{agg}} + \mathrm{Var}_{\mathrm{ind}}}. \quad (6.7)$$

Here $\mathrm{Var}_{\mathrm{ind}}$ shows the individual-level variance. $\mathrm{Var}_{\mathrm{agg}}$ shows the adjusted variance of the city aggregate rate.[5]

---

[3] $\overline{f(X)}$ denotes the mean level of the function $f(X)$; $\overline{f(X)}_j$ denotes the mean level of the function $f(X)$ in city j.

[4] "[I]n the equilibrium of this model two agents who are separated by K other agents will … do exactly the same thing if there are no fixed agents between them" (Glaeser and Scheinkman 2001, p. 95).

[5] We set

$$\mathrm{Var}_{\mathrm{agg}} = \mathrm{Var}\left[\frac{1}{\sqrt{N}} \sum_{i=1}^{N} \left(A(i) - \frac{\overline{f(X)}_j}{1-\alpha}\right)\right].$$

6.2.3 Price Equation in Henrich's Context

"The Price equation is a simple statistical statement" on the expected change of the frequency of a gene x. In this gene, x_i denotes an altruistic allele if individual i retains $x_i = 1$, while an egoistic allele if $x_i = 0$. x_i can then express the current frequency of this gene. Let w_i be the number of offsprings of i, namely, the absolute fitness of i while $\bar{w}$ the average fitness of the population. In Henrich's contribution, "we ignore any effects arising from the transmission process (e.g. recombination, mutation, etc.)".[6] (Henrich's drafts, p. 18). It then follows the Henrich equation (6.8):

$$\bar{w}\Delta\bar{x} = \underbrace{\beta_{w_j x_j}\,\mathrm{Var}(x_j)}_{\text{selection between-groups}} + \underbrace{E(\beta_{w_j x_{ij}}\,\mathrm{Var}(x_{ij}))}_{\text{selection within-group}}. \tag{6.8}$$

Here $\beta_{w_{ij}x_{ij}}$ means the within-group regression coefficients of w_{ij} with respect to x_{ij} in group j. This type of Price equation "tells us that the change in the frequency of allele created by natural selection acting on individuals can be partitioned into between-groups and within-group components" (Henrich's drafts, p. 18).

6.2.4 Difficulties of Price equation

It is easily seen that our Price equation shares the same idea as the simple model of Glaeser and Scheinkman (2001). In those kinds simply a statistical relationship is expressed to measure a social interaction by decomposing both in between-groups effects and in within-group effects. This way is quite rightful in practice in measuring contributions of interactive factors due to between-groups and within-group as shown generally in a form of:

$$\bar{w}\Delta\bar{x} = \underbrace{\mathrm{Cov}(w_j, x_j)}_{\text{selection between-groups}} + \underbrace{E_j(\mathrm{Cov}(w_{ij}, x_{ij}) + E(w_{ij}\Delta x_{ij}))}_{\text{selection and transmission within-group}}. \tag{6.9}$$

But this kind of formulation does not entirely mean to bring an overarching mechanism working between between-groups and within-group. Without elucidating the overarching mechanism, the signs of between-groups/within-group effects cannot be definitely ascertained.

> [S]election on group level, as a rule, is slow and much less effective than on the level of individual (Hofbauer and Sigmund 1989, p. 109).

The assumption of a mutant gene to promote altruistic behavior for his group may be rejected by the reason that it reduces the fitness of its carrier. In order to

[6] In the Glaeser–Scheinkman model, no sorting occurs within city. Cities may be replaced with groups in the Price equation. $A_i - \overline{A}_j$ may be x_{ij}.

make group selection well workable, transmission mechanism is indispensable, as Henrich states. Cultural group selection based on such a transmission could be viable if and only if a certain macroscopic order should be implemented in the interactive field. In other words, a promising way to take a cooperative system formation into account is to introduce a macroscopic order in the concerned field in which individual agents act bilaterally with their other groups. It may be the key issue to find such a binding condition in group selection to form a macroscopic order. Putting another way, we must look for a macroscopic microeconomic linkage (i.e. an overarching mechanism to illustrate the processes of a disperse social system).

In the next section, we can only refer to the binding conditions for a macroscopic order in the theory of consumer demand, although it is not yet a binding condition on covariances of group selection for cooperation.

6.3 A Digression: An Overarching Mechanism Working in Market Demand

6.3.1 Slutsky Equation on Price Changes

Some digression may be required for a while. It is very instructive to learn how a macroscopic binding condition works to organize a kind of economic law. In the theory of consumer demand, we have a well-known equation called Slutsky equation on demand variation due to the changes of price p. $f^i(p, x^i)$ denotes a demand function of individual $i \in I$ who belongs to the cloud (income class) x^i. The Slutsky equation system requires the following hypothesis in order to derive the law of demand.

Hypothesis (i). The average income effect matrix A is positive semi-definite.

Hildenbrand then proved quite ingeniously that Hypothesis (i) can be equivalent to Hypothesis (ii), provided the demand functions satisfy the budget identity (see Hildenbrand 1994, Appendix 5).

Hypothesis (ii). [Increasing spread and expanding dispersion of household's demand.] For every sufficiently small $\Delta > 0$, the cloud $\{f^i(p, x^i + \Delta)\}_{i \in I}$ is more spread than $\{f^i(p, x^i)\}_{i \in I}$. In other words, the second moments of the former are greater than those of the latter.

Dispersion on the clouds can be defined to measure heterogeneity of households. To remove the so-called Giffen effects, the average income effect matrix A must be positive semi-definite. According to a new theory of Hildenbrand, this corresponds to the fact that "dispersion in each income to measure heterogeneity of households may increase, as income size increases". In short, it then holds:

$$\text{Ell}(\text{Cov } v(x^i + \Delta)) \supset \text{Ell}(\text{Cov } v(x^i)). \tag{6.10}$$

Here Ell means ellipsoid of dispersion in the covariance matrix Cov v. Our empirical test thus comes down to calculate the matrix of covariance around the mean on the clouds.

6.3.2 A Macroscopic Microeconomics

As Hildenbrand (1994) smartly proved, thus, it is verified that individualistic demand behavior could be well-defined to make income effects always positive, provided that macroscopic variables like variances and covariances on spending among consumer goods are pertinently bound under a certain set of hypotheses.[7] In a sense that a microeconomic behavior on demand could be solved by introducing a macroscopic condition, Hildenbrand called this approach "macroscopic microeconomic approach". (Hildenbrand 1994, p. 74).

Hildenbrand in cooperation with Härdle (1990) working in the field of non-parametric testing, in fact, estimated the matrices of second moments as well as covariances using the available data like *UK Family Expenditure Survey* (FES), *French Enquâte Budget de Famille* (*EBF*) and the surveys in other developed countries but except for Japan.[8]

Now we return our main topics. It seems us difficult to find a binding condition on group selection towards a cooperative system in the case of altruistic and egoistics alleles. First of all, we must be keen to promote some intensive efforts to achieve such empirical analysis as done in Glaeser and Scheinkman (2001).

References

Aoki M (1996) New approaches to macroeconomic modeling: evolutionary stochastic dynamics, multiple equilibria, and externalities as field effects. Cambridge University Press, Cambridge, UK

Arthur WB (1994) Increasing returns and path dependence in the economy. University of Michigan Press, Ann Arbor

Aruka Y (2001) The law of consumer demand in Japan: a macroscopic microeconomic view. In: Takayasu H (ed) Empirical science of financial fluctuations: the advent of econophysics. Springer, Tokyo, pp 294–303

Durlauf SN (1997) Statistical mechanics approaches to socioeconomic behavior. In: Arthur WB, Durlauf SN, Lane DA (eds) The economy as an evolving complex system II. Addison-Wesley, Reading, MA, pp 81–104

[7]Subgroups may be interpreted as income classes in the society; fitness of the subgroup may be measured in terms of spending on consumer goods.

[8]According to the above method Aruka (2001) estimated the matrix of second moment, as well as the matrix of covariance in the family expenditure data in Japan of the period 1979–1998.

Durlauf SN (2000) A Framework for the Study of Individual Behavior and Social Interaction. SSRI Working Paper Series. Social Systems Research Institute of the Economic Department, The University of Wisconsin, Madison
Föllmer H (1974) Random economies with many interacting agents. J Math Econ 1:51–62
Friedman D, Sighn N (2001) Evolution and negative reciprocity. In: Aruka Y (ed) Evolutionary controversies in economics: a new transdisciplinary approach. Springer, Tokyo, pp 103–114
Glaeser EL, Scheinkman JA (2001) Measuring social interactions. In: Durlauf SN, Young HP (eds) Social dynamics. Brooking Institute, Washington, DC, pp 83–131
Härdle W (1990) Applied nonparametric regression. Cambridge University Press, Cambridge, UK
Helbing D (1995) Quantitative sociodynamics: stochastic methods and models of social interaction processes. Kluwer Academic, Dordrecht
Hildenbrand W (1971) Random preferences and equilibrium analysis. J Econ Theory 3:414–429
Hildenbrand W (1994) Market demand. Princeton University Press, Princeton
Hofbauer J, Sigmund K (1989) The theory of evolution and dynamical systems. Cambridge University Press, Cambridge, UK
Weidlich W, Haag G (1983) Quantitative sociology. Springer, Heidelberg

Part II
Moral Science of Heterogeneous Economic Interaction

Chapter 7
Exploring the Limitations of Utilitarian Epistemology to Economic Science in View of Interacting Heterogeneity

Yuji Aruka

7.1 Suggesting a New Art of Life

The principles of political economy born before utilitarianism seized power in economics were entirely irrelevant to the kind of utility maximization. Utilitarianism made economists share a unique definite purpose for the art of life, thus becoming to play the crucial role in arguing economics almost everywhere. We can easily find our main prototype of modern economic ideas from the classical source of literatures of utilitarianism, in particular, James Mill who suggested the "Art of Life", whose ultimate end is *happiness* in the society. Put another way, utilitarianism is a kind of art which has ultimately recourse to the sole value judgment on happiness either personally or interpersonally.

According to Hare (1963) and Riley (1988, p. 156), there is no logical link between a descriptive premise, by itself, and a prescriptive conclusion. It becomes possible to draw a logical inference if a prescriptive premise is added. Bentham and Mill called the prescriptive premise "springs of action". Mill believed in the analogy between a standard of value and a divine rule. In this sense, Mill's rule to apply a value of standard must be uniquely single to guide to conduct in all the department of the Art of Life. Sen (1970, p. 59) did restrict the domain for such a value standard to the non-basic judgment. In Mill,

> science carries implications for our basic value judgment per se.

The art of life requires *practical knowledge*. Practical knowledge is the collection of rules which summarize the means by which happiness, the end of the art is attained. By practical knowledge, *is* could be transformed into *ought*. Many

---

Reprinted from Annals of the Japan Association for Philosophy of Science 13(1), Aruka, Y., Exploring the Limitations of Utilitarian Epistemology to Economic Science in View of Interacting Heterogeneity, 27–44 (2004). With kind permission from Japan Association for Philosophy of Science, Tokyo.

Y. Aruka
Faculty of Commerce, Chuo University, Higashinakano, Hachioji, Tokyo 192–0393, Japan
e-mail: aruka@tamacc.chuo-u.ac.jp

DOI 10.1007/978-3-7908-2618-0_7, © Springer-Verlag Berlin Heidelberg 2011

utilitarian always think of the link between science and utility in terms of "is" and "ought". In summary,

> *the maximization of utility*, what *ought* to be, is, for Mill, contingent on what *is*. (Riley 1988, p. 134)

We have a helpful book of Jonathan Riley (1988) on James Mill's utilitarianism where Riley thoroughly inspected a fundamental utilitarian set of distilled ideas. By the use of this book, for a while, we could provide the readers with a compact brief of the whole foundation of utilitarianism. Many phrases in this section are often cited or partially recast from the original statements in Riley (1988, Chap. 7), although the ordering of phrases is arranged by the present author's idea.

In the first section of this article, we summarize *the construe of epistemology* of utilitarian economics as the art of life. In the second section we explore *the failures of homogeneous agent* of economic rationality, in particular, as for the interpersonal comparison. We then suggest *a new art of life* to understand the reality by constructing a macroscopic microeconomics.

7.2 The Epistemology of Utilitarian Economics

7.2.1 The Art of Life as the Guide of Action and Science

Mill argues the guide of action, i.e., the Art of the Life, in combination with the law of causation.

The guide of all our action "The Law of Cause and Effect" in its objective aspect, as the fundamental principle in the order of universe, [is] the basis of most our knowledge, *and the guide of all our action.*

The law of causation is "an empirical law coextensive with all human experience; at which point the distinction between empirical laws and laws of nature vanishes".

According to Riley (1988, p. 157), the three steps could be required to achieve our action in the society.

The first step On the scientific view of nature, any human action is related to a series of antecedents thought uniform laws, and the series of antecedents itself depends on all the other chains of causation in Nature.

The second step Any intention or volition is liable to be frustrated by unforeseen changes in some other phenomena which have their impact through the interconnected chains of causation.

The final step In Mill's argument for an ultimate standard of value given the complex interdependence which characterizes human phenomena, it becomes necessary to have a single principle to "umpire" between conflicting ends.

In our article, we will focus on the second step *in terms of heterogeneous interacting agents* to argue a possibility of *a new art of life*.

7.2.2 *The Art of Life*

Science and Art: In the above sense, economics in view of utilitarianism is not merely science but also belongs to the art of life on human nature. According to Mill:

By "science", he apparently means *theoretical knowledge* of the actual world, as it *is*. Scientific knowledge [including the psychological theorems] can be expressed in the form of theorems or "laws" which summarize *uniformities* among the phenomena of nature (including human nature).

By "art", to make the world what he thinks it *should* be. *Practical knowledge* can be expressed in the form of rules which summarize the means by which the ends or goals of art are to be attained in the actual world.

7.2.2.1 Art: A Priori Knowledge – Self-Evidence

In Mill, it holds the following relations.

Logic is also an art which has a practical end, i.e., "truth". At least in part, then, the goal of the *art* of logic is to ascertain scientific knowledge, including the psychological theorems which comprise the *science* of logic itself (Riley 1988, p. 134).

A priori knowledge Mankind *judged* evidence, and often correctly, before logic was a science, or they never could have made it one. It seems, then, that our practical knowledge of what the process of reasoning *should* do must, at least in part, *precede* our scientific knowledge of what the process of reasoning *is*. . . .

Self-evident We have some knowledge of the practical *end* of reasoning before any reasoning takes place. In other words, some part of "truth" is *self-evident* to our consciousness, and correct inferences must be compatible with such evidence.[1]

7.2.2.2 Conviction: Judgment Before Reasoning

Conviction or belief [Man] experiences such mental phenomena, is what is meant by the direct knowledge of his consciousness. According to Mill, this conviction or belief is *the most certain knowledge* a person can have. Moreover, as "our model of certainty", the direct knowledge we have of our own mental states is "the test to which we bring all other convictions" (Riley 1988, p. 136).

Judgment If it is the case, we are able to *judge* of at least some evidence (i.e., the existence of our own mental phenomena) *before* we know what the act of reasoning is.

[1]I know directly, and of my own knowledge, that I was vexed yesterday, or that I am hungry today. The knowledge that we have such and such an emotion . . . precedes our reasoning about the mental phenomena in question (Riley 1988, p. 35).

7.2.2.3 From Direct Knowledge to Scientific Knowledge

Now the way how to obtain *the most certain knowledge* can be argued.

The direct knowledge due to eyesight, or direct intuition, may contain a kind of mistake. The knowledge rather comprises of crude information. To make more precise information, we must engage in reasoning, using the crude truths of consciousness as "starting points" (Riley 1988, p. 137).

Nature exists in simultaneity and succession whose phenomena both are not necessarily from mathematical inferences alone. A phenomenon of succession indicates scientific investigation. Simultaneity or causation does not imply merely the invariable conjunction of antecedent and consequent means.

Scientific knowledge is contingent on experience.

Relative certainty In matter of evidence, scientific knowledge never admits of absolute certainty.

Inducibility So the logic of experience (truth) views all inference as *inductive.*

Remark on cultivation of tastes by knowledge accumulation Individualistic tastes may be *cultivated* as *knowledge* of an individual or the society to which an individual belongs *accumulates.* In order to attain the state of "mind-in-itself", this point of view had to be dropped from economics. Otherwise, the engine for *the evolution of utility function* must be ignited in the earlier times.

7.2.2.4 Mind-in-Itself

[Mill] concedes that he believes in a permanent identity, *a "thread of consciousness"* which transcends the succession of mental phenomena associated with an individual's different bodily states (Riley 1988, p. 142). In the event, Mill conjectures the nature of memory-in-itself. In any event,

Mill's epistemology is not affected by the failure of his psychological approach to give an adequate account of our belief in a permanent personal identity. The belief in personal identity may be accepted as an original "truth of consciousness" a starting point from which reasoning proceeds.

The "mind-in-itself" [Mill] can and does not ignore not only the problem of personal identity, but also the distinction between belief and imagination, treating it as an "ultimate fact" about the "mind-in-itself", beyond psychological explanation (Riley 1988, p. 142).

7.2.2.5 Preservation of Intrinsic Rationality as the Mind-in-Itself

In economics, *the mind-in-itself* may be replaced with *intrinsic rationality.* Its feasibility could be guaranteed by *free will.* Some awkward *ethological* reasons can interfere individualistic rationality. Intrinsic rationality in economics, as long as it is part of the mind-in-itself, never is tested by experiments.[2]

In economics, the properties of *intrinsic rationality* for consumer preference usually is as follows:

1. Reflexivity
2. Completeness
3. Transitivity
4. Continuity
5. Monotonicity
6. Convexity

It must be noticed that *the first* 1–4 *properties* are necessary to guarantee utility function, and some weak convexity on preference relation is not necessary for it.[3] In this view, the so-called indifference map of textbooks on microeconomics is not indispensable for the existence of utility function.

Mathematical economics in the twentieth century was quite successful to discover *isomorphism* between the preference relation and the utility function. This success was really innovative, as long as "intrinsic rationality" seems universe. The utility function pertinently defined is just equivalent to the preference ordering on the domain of multiple commodities. *Utility function* in economics can usually give each unique value to each set of commodity basket through the affine transformation respectively. These properties guarantee the possible preference relations the maximal elements to guarantee the utility function (Table 7.1).

Table 7.1 Equivalence between preference relation and utility function

properties	$\xrightarrow{indifference-map}$	preference ordering
$isomorphism \downarrow$		$isomorphism \downarrow$
utility function	$\xrightarrow{quasi-concavity}$	utility maximization

7.2.3 *The Art of Human Nature for Mental Phenomena*

Here we use the terms in the following sense that

Concept is built up by a succession of judgments.

Judgment (=*decision*) is an act of will which affirms some relation between attributes.

[2]Experimental economists who anticipate the derivation of rational choice axioms from conducting human experiments.

[3]See for example, Hildenbrand and Kirman (1988).

Reasoning (=*inference*) is the act of will by which the mind makes judgments "not intuitively evident".

Intention (=*volition*) when we are said to intend the consequences of our actions, means the foresight, or expectation of those consequences.

7.2.3.1 The Five Classes of Mental Phenomena

In the science of human nature, we have *the five classes of mental phenomena:* sensations, thoughts or ideas, emotions, volitions, and desires (and aversions) (Riley 1988, p. 145).

1. *Sensations* e.g., touch, sight, pleasure, pain in the class of feeling.
2. *Thoughts or ideas* is mental impressions or representations of sensations.
3. *Emotion* is constituted by simple emotion (e.g., fear, wonder) and complex emotions (e.g., aesthetic feelings, moral feelings).
4. *Volitions as Intentions* is an intention to act, an action being the combination of an intention and its physical consequences.
5. *Desire as Motives* is a seeking after pleasure, and an aversion is a seeking to avoid pain. "A desire is more than merely an idea of pleasure: it also involves a reference to *action*".

7.2.3.2 Choice Under Preference

If a person has conflicting motives, then the strongest among them in relation to his ideas of pleasure and pain, determines his volition. A person's strongest motive, his strongest present desire or aversion, can be termed his *preference* (Riley 1988, p. 149).

Preference is some belief by which an agent *expects* the probable success. This is a causal antecedent of action.

Action Volitions, whether caused by present desire or by force of habit, themselves cause *actions*. Mill regards an action to be the combination of volition and its physical consequences or effects. All action is voluntary or intentional.

Choice Actions constitute *choice* only if the agent's volitions are determined by his own thinking (Riley 1988, p. 150).

Custom A person whose volitions are dictated for him *by custom*, without any conscious act of thought on his part, does not make *choice*.

7.2.3.3 The Proposition of Actions Chosen

At this stage, the choice for *the representative agent* is *implicitly* defined. The representative agent *neutralizes* the effects of interaction among the various agents.

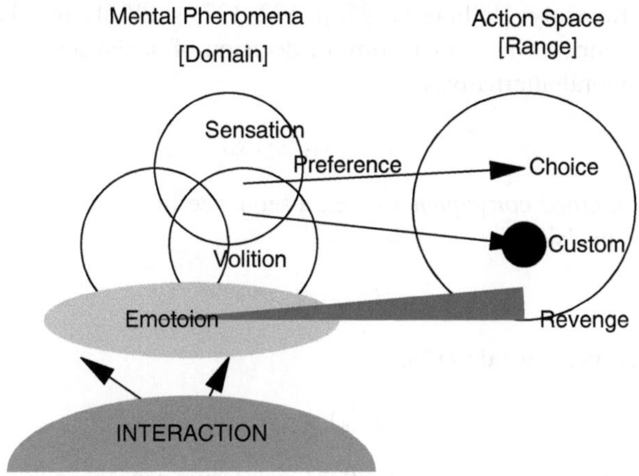

Fig. 7.1 Choice space in the representative agent

Behaviour constitutes *action*, if and only if action is preceded by *desire* combined with *some knowledge of the (probable) consequences of the behavior in question*. Each agent will require *a common knowledge* as long as the representative agent is supposed to be *admissible*.

Common knowledge can be derived from the properties on preference.

Remarks on expected utility theory In Fig. 7.1, suppose the Action Space to replace with the Event Space. In the theory of expected utility, probable consequences can be assigned by probability distribution on the concerned set of events.

7.2.4 An Alternative Formulation of Decision

Mill, on the other hand, doubts that we have any a priori knowledge of external reality (including other minds). The need for the a priori approach as long as any existing metaphysical belief about our own minds or about external reality can be explained by psychological theorems (Riley 1988, p. 143).[4]

Social interaction of agents is a process of mutual influence for better or worse. Utility of each agent is always affected not only by others' behavior but also by subgroup's characteristics to which each belongs. Decisions in many cases *evolve* as a result of *nonlinear aggregate dynamics*, which can be expressed in terms of the *master equation*, as long as the variables in a system are a few. The key issue to achieve this hinges on the definition of "the transition rate". Social dynamics of Weidlich and Haag (1983), in other words, *social synergetics* is a pioneering work

[4]Although theory is required to interpret the facts of our experience, it does not follow that our minds create the theory a priori, before we see the facts.

in this sense. Recently, Helbing (1995, p. 132–137) skillfully in this context was successful to define an analytical form of decision of social interaction. Suppose there are behavioral alternatives:

$$x_1, \cdots, x_n, \cdots, x_s.$$

We have *imagined consequences*, i.e., a sequence of consequence anticipation, of the possible decisions:

$$i_0, i_1, \cdots, i_n, \cdots.$$

When i_n favors x_n, we describe

$$f(i_n) = x_n.$$

Thus *decision* on x has been made after k successive consequence anticipation on the same x, i.e.,

$$\left.\begin{array}{ccc} i_n & \rightarrow & x \\ \vdots & & \vdots \\ i_{n+k} & \rightarrow & x \end{array}\right\} \text{The same consequences to be anticipated}$$

Helbing has the insight that utility has a past history of decision sequence and also cannot be independent from it. This utility function should rather be called *a psychologically complex value*. Consequently, utility will fluctuate like in human mind. Then he defines *utility* or *preference* of decision on alternative x for agent α who has finally made decision y:

$$U_\alpha(x|\, y; t).$$

The probability for occurrence of i, due to $f(i) = x$, will be

$$p_\alpha(x|\, y; t).$$

The probability is supposed to increase as the preference $U_\alpha(x|\, y; t)$ increases. We by virtue of the multinomial logit model in the theory of discrete choice (see, e. g., Ben-Akiva and Lerman 1985, Chap. 5) normalize these probabilities to assure nonnegativeness:

$$p_\alpha(x|\, y; t) = \frac{e^{U_\alpha(x|\, y; t)}}{\sum_{x'} e^{U_\alpha(x'|\, y; t)}}.$$

The probability that the next k arguments are all in favour of alternative x is described by the use of $\hat{p}_\alpha$ as follows:

$$\hat{p}'_\alpha(x|\, y; t) = [\hat{p}_\alpha(x|\, y; t)]^k = e^{k\hat{U}_\alpha(x|\, y; t)},$$

where the scaled preferences is

$$\hat{U}_\alpha(x|\,y;t) := U_\alpha(x|\,y;t) - \ln \sum_{x'} e^{kU_\alpha(x'|\,y;t)}.$$

The *attractiveness* of decision x for agent α who chose behavior y will be expressed as:

$$U'_\alpha(x|\,y;t) := k\hat{U}_\alpha(x|\,y;t) - \ln \sum_{x'} e^{k\hat{U}_\alpha(x'|\,y;t)}.$$

By the use of $U_\alpha(x,t)$, utility of behavior x, and the concept of distance between behaviors x' and x, Helbing concludes *readiness*:

$$R_\alpha(x'|\,x;t) := \frac{e^{U_\alpha(x',t)-U_\alpha(x,t)}}{D_\alpha(x',x|\,t)}$$

He specified the following elements for decision on x by subgroup α:

- *Flexibility* at time t: $\nu_\alpha(t)$
- *Distance* between x and x' at time t: $D_\alpha(x',x;t) = D_\alpha(x,x';t) > 0$
- *Effort* at time t: $S_\alpha(x',x;t) := \ln D_\alpha(x',x;t)$

Thus the social transition rate for the subgroup α, w_α could be summarized into

$$w_\alpha(x'|\,x;t) = \nu(t)R_\alpha(x'|\,x;t)$$

for $x' \neq x$. Thus such a social transition rate will be operated in the master equation of social process.

7.3 Some Failures of the Representative Agent in View of Interacting Heterogeneity

7.3.1 The Foundation of Interpersonal Comparison

According to Riley (1988, p. 153), interpersonal comparison could be comprised of the following two stages i.e., introspection and inference:

The first stage is the *introspection,* whereby a person observes the sequences of his own mental phenomena which follow physical modifications of his body, and precede the performance of his actions. Recall that any action has physical effects, i.e., actions change physical circumstances and thereby cause a person to experience new sensations and new successions of feelings.

The second stage is *the inference*, whereby a person generalizes his personal knowledge of his own mental phenomena to other persons when they experience the same kind of bodily modifications and perform the same kind of actions. In other words, if he observes any two persons (one of whom may be himself) experiencing bodily modifications of the same class and performing actions of the same class, then he infers the existence of the same succession of feelings in both persons.

Rather than extending sympathy, he is inferring that the other person is actually experiencing the same kind of desires which he himself is experiencing in similar circumstances, despite any subjective disagreement which might nevertheless be reported.

7.3.2 *Transitivity of Inference for Interpersonal Comparison*

Let a series of facts connected by uniform sequence to be

$$s = \{\text{beginning } b, \text{middle } m, \text{end } e\},$$

where

the beginning (b) is modifications of my body,
the middle (m) is feelings,
the end (e) is outward demeanour.

The sequence s means *transitivity* among s:

$$\text{if } b \rightarrow m \text{ and } m \rightarrow e, \text{then } b \rightarrow e.$$

1. In my own case I know that *the first link b* produces *the last e* through *the intermediate link m*, and could not produce it without m.
2. Experience, therefore, obliges me to conclude that there must be *an intermediate link m*.
3. Experience must either be the same in others as in myself, or a different one.[5]

7.3.2.1 Intransitivity Due to the Heterogeneous Agents

In order that such an inference could be admissible, there would be required that all the agents appearing in such a process all are homogeneous. If the agents there were heterogeneous, such an inference could be break down. Take, for instance, the case that there are two different types of agent *either* with (a) *the same phenotype but the different genotype or* (b) *a common genotype with two different phenotypes* (Table 7.2).

---

[5]I must either believe them to be alive, or to be automatons: and by believing them to be alive, that is, by supposing the link to be of the same nature as in the case of which I have experience, and which is in all other respects similar, I bring other human beings, as phenomena, under the same generalizations which I know by experience to be the true theory of my own experience. Cited from J. S. Mill (1865) by Riley (1988, p. 154).

Table 7.2 Heterogeneity, hierarchy, and overlapping of alternatives

Sub-group			(1)					(2)					(3)		
Genotype			α			α				β			β		
		↙		↘			↖		↗			↙		↘	
Phenotype (Alternatives)	A				B			A			B				C

There always appear the heterogeneous agents in our society. Heterogeneous agents usually are distinguished from their different preferences. As long as preference is similar among agents, all the agents can be homogenous. In economics, however, heterogeneity does not necessarily come from the human nature in itself. Heterogeneity can come into being among agents in the form of inequality of either income or wealth. The economic effects thus are constituted by both the supposed intrinsic properties and the macroscopic orders of economic variables.

A typical example is the law of demand when we deal with the *Slutsky equation* on consumer demand, one of the most fundamental equations in price theory. This procedure will be dealt with in the final section.

7.3.2.2 Metonymy in Economics and Intransitivity of the Preference Links

"Metonymy" is defined as the "figure of speech that consist in using the name of one thing for that of something else with which it is associated". Hildenbrand (1994, p. 25) illustrated *metonymy in economics* by the use of the following example.

> Consider a population consisting only of one person households that are either male or female. All males in the population have the same demand function $f^{♂}$ and all females have the same demand function $f^{♀}$. Let $\pi(x)$ denote the percentage of males among the x-households[household with the income level x]. If the function $\pi(x)$ depends on x, for example, if it is an increasing function, then the entire population might not be *metonymic*. Indeed, the distribution of x-households' demand functions is not independent of the income level x, since this distribution gives the weight $\pi(x)$ to the demand function $f^{♂}$ and the weight $1 - \pi(x)$ to the demand function $f^{♀}$.

For example, the aggregate demand for cloth at the human level can be divided into the demands at the different genders level, indeed. To a degree of heterogeneity to the extent that observer *zooms in*, the demand for cloth must be *hierarchical*. The introduction of hierarchy implies another relationship to be added by which uncertainty could be brought up. Only if *the representative agent* were *ubiquitous* all in the layers of composition, metonymy could just be applied. The introduction of gender may engender the rule of transitivity.

Suppose the intermediate link m to be "generating preference function in economics". The demand to prefer a cloth can be generated by gender.

1. Gender is the first link.
2. The first link passes information to the second link: Female preference map.
3. The third link is an outcome of the second link.

The second link must be unique and stable, giving a *probable consequence*. In this case, the demand is considered just *metonymic* on the demand. When a series of metonymy does not hold, transitivity of inference does not hold. Without transitivity of inference, we could not attain the same final link.

7.3.2.3 A Nonlinear Preference Tree Structure and Intransitivity

In the restaurant, suppose that the customer is faced with two different menu courses of two dishes. Let a course be a composite commodity of two different dishes. There are assumed to be four single dishes *A, B, C, D*. Course *A* is a set $\{A,B\}$ denoted by α, and Course *B* a set $\{C,D\}$ denoted by β. α is an aggregate of metonymy of the dishes $\{A,B\}$, β an aggregate of $\{C,D\}$.

It might happen for an agent the case that

$$A \preceq C, A \preceq D, B \preceq C, B \preceq D,$$

while

$$\alpha \succeq \beta.$$

In our aggregation of the dishes in the restaurant, the preference directions might never be preserved between the elements of the upper layer and the elements of the lower layer. The so-called "nonlinear" aggregation may not hold *metonymic* (Table 7.3).

Remark on the case of traditional expected utility. In the theory of expected utility, it is noticed that the single dishes A, B, C, D are the events. If each event were assigned each probability given rise to the consequences, those probability distributions thus obtained could be considered the lotteries. In economics, for simplicity, the compound lotteries could always be reduced into the simple lotteries. This treatment is a kind of *metonymy*.

7.3.3 *The Law of Demand Under a Macroscopic Order*

In the elementary course of microeconomics, *interpersonal comparison* usually is shunned. Instead of referring to interpersonal comparison, economists are fond of comparison among the commodity baskets. It was a kind of wisdom to come

Table 7.3 Nonlinearity under metonymy

Composite commodity			α			$\succeq$			β		
Aggregation		$\nearrow$		$\nwarrow$				$\nearrow$		$\nwarrow$	
Single commodity	$(A$				$B)$	$\preceq$	$(C$				$D)$
	A	$\preceq$	C, D		B	$\preceq$	C, D				

through the difficulties of interpersonal comparison. Economists had an easy expectation to formulate the aggregate demand with resort to the individualistic preference relationship. They however were faced with a grave problem of failure to achieve this expectation.

The hypothesis of individualistic utility maximization implies a nonpositive Slutsky substitution. In the event, some basic property in an individualistic demand cannot be preserved in the aggregate demand, i.e., the market demand. At the beginning of the so-called *weak axiom of revealed preference* suggested by Samuelson (1938), we were vexed by a counterexample.

In a microeconomic view, our indifference map on the commodity basket comparison always intrinsically assures a decrease of demand on the concerned commodity of a relative price increase to the compared commodity. We call this a nonpositive Slutsky substitution effect (=weak axiom of revealed preference). We denote a demand f^i of household i to consume the basket z by f^i when the price vector p and income x is given:

$$f^i(p,x) := \arg \max_{pz \leqslant x} u^i(z).$$

Here $\partial^2 u^i(z)$ indicates the matrix of second derivatives of the utility function u^i (Hildenbrand 1994, p. 18).

As long as the hypothesis of individualistic utility maximization implies a nonpositive Slutsky substitution effect, the demand thus is expected to be monotonically decreased if the price is continuously increased. Vice versa. We call this reaction *the law of demand.*

If we employ individualistic utility maximization to derive the law of demand, we should be forced to restrict the domain of the utility function adding to *the condition of* Mitjushcin and Polterovich (1978):

$$-\frac{z\partial^2 u^i(z)z}{z\partial u^i(z)} < 4.$$

A well-known obstruction against the law of demand in general is due to the *Giffen effect* which means to offset the *substitution effect* by the *income effect* when a price change on commodity happened. Given continuously a price change, the positiveness of income effect therefore is not always assured.

Remark on vulnerability of the substitution effect. So far in economics, substitution effect is regarded as robust. It however is important to know how *vulnerable* the substitution effect might be in terms of the condition of Mitjuschin and Polterovich. We rather had restricted the domain of preference to a class of *volitions* in Fig. 7.1.

7.3.4 Introduction of a Macroscopic Order

Hildenbrand (1994) has ingeniously solved the grave problem of Slutsky equation by introducing *a microscopic order*. There are two ways for our solution. Both were given by Hildenbrand (1983, 1994), respectively.

7.3.4.1 The Law of Demand and the Pareto Law

Hildenbrand (1983) proved *the law of demand* in connection with *the Pareto law.*

Theorem *(Hildenbrand 1983, p. 1003). All individuals of a subgroup x-household have the same consumption behavior described by a common demand function $f(p,x)$, which is assumed to satisfy the weak axiom of revealed preference. Suppose that the distribution of individual total expenditure x is given by a decreasing density ρ with $\rho(0)>0$, i.e., the Pareto law. It then holds the law of demand that the market demand function*

$$F(p) = \int f(p,x)\rho(x)\mathrm{d}x$$

is monotone, i.e., for any two price vectors p and q has

$$(q-p)(F(q)-F(p))\leqslant 0.$$

The Pareto law may be interpreted as a decreasing density on income size. Here we identify the expenditure with income. As the income class is raised, the expenditure is supposed to be risen while the density of income–expenditure class is decreasing.

By the above theorem, it turns out that *the law of income distribution* controlled the income effects generated by the interacting activities in the market.

For a long time, we always were vexed by the following kind of wonder on economics:

> [F]or a very long time *the Pareto law* has lumbered the economic scene like an erratic block on the landscape – unconnected with anything else in the field. This justifies an old epigram of Professor Kalecki ...: "Economics consists of theoretical laws which nobody has verified and of empirical laws which nobody can explain". (Steindl 1965, p. 18)

We thus just prepared for an entire integration of positive theory and empirical analysis in economics.

7.3.4.2 Macroscopic Microeconomics

In the same context, we have a more elegant way of implementing a macroscopic order. Macroscopic variables like *variances* and *covariances* on spending among consumer goods are pertinently bound under a certain set of hypotheses.[6] A microeconomic behavior on demand could be solved by introducing a macroscopic

[6]Subgroups may be interpreted as income classes in the society; fitness of the subgroup may be measured in terms of spending on consumer goods. In the context of evolutionary economics, on the other hand, we can use variances and covariances to argue cultural group selection,

condition, Hildenbrand called this approach "macroscopic microeconomic approach" (Hildenbrand 1994, p. 74).

To remove the so-called Giffen effects, the average income effect matrix A must be positive semi-definite. This corresponds to the fact that "dispersion in each income to measure *heterogeneity of households* may increase, as income size increases". In short, it then holds

$$\mathrm{Ell}(\mathrm{cov}\ v(x^i + \Delta)) \supset \mathrm{Ell}(\mathrm{cov}\ v(x^i)). \tag{7.1}$$

Here Ell means ellipsoid of dispersion in the covariance matrix cov v. Our empirical test thus comes down to calculate the matrix of covariance around the mean on the clouds.

Hypothesis (i). *The average income effect matrix* A *is positive semidefinite.*

Hypothesis (ii). [Increasing spread and expanding dispersion of household's demand] *For every sufficiently small* $\Delta > 0$, *the cloud* $\{f^i(p, x^i + \Delta)\}_{i \in I}$ *is more spread than* $\{f^i(p, x^i)\}_{i \in I}$. *In other words, the second moments of the former are greater than those of the latter.*

Hildenbrand then proved quite ingeniously that *Hypothesis (i)* can be equivalent to *Hypothesis (ii)*, provided the demand functions satisfy the budget identity. (See Hildenbrand 1994, Appendix 5).

It is to be noted that *dispersion* on the clouds can be defined to measure *heterogeneity of households.*

7.4 Concluding Remarks

We investigated the basic properties of utilitarian economics whose domain is excessively limited to *the class of volition* and omitting the other important aspects of human behavior. As based on this construe of utilitarian economics, we shall make the utility function *evolute* by extending the field of domain, indeed. We really need *the evolution of utility function.* In the final subsections, we have looked around some ways to introduce a macroscopic order in microeconomics. It is quite instructive that microeconomic properties have been revealed in association with some macroscopic variables. Without any filed of macroscopic order, we cannot refer to some genuine microscopic properties. Economics has ever been irrelevant to such a concept of temperature.[7]

coevolutionary processes, and large-scale cooperation in line with George Price's equation (Price 1972) in biology. See Aruka (2004).

[7]The author adopted the same citations from Hoover (2001) and Weidlich (2000), and also in Aruka (2003, p. 271) in order to emphasize the importance of macroscopic microeconomics.

Hoover (2001, pp. 74–75) states:

The physicist who has successfully reduced the ideal gas laws to the kinetic theory of gases does not then abandon the language of pressure, temperature, and volume then working with gases or try to use momentum, mass, and velocity as the principal phenomenological categories for discussing the macroscopic behavior of gases. But *economists have taken a different track.*

Similarly, Weidlich (2000, p. 11) stated concisely:

In the case of *physical systems* the introduction of the field concept proved to be most important for the deeper understanding of the systemic character. Take the characteristic example of electromagnetic interaction: Particles, the elements of the physical system, possess – besides other properties – both mass and electric charge, and generate in their environment a gravitational as well as an electromagnetic field. The field contributions of many particles are *superposed* and form a *collective field.*

Macroscopic microeconomics as Hildenbrand (1994) suggested provides us with an important hint of creating *a new art of life.*

References

Aruka Y (2003) The complex adaptive processes in economics by heterogeneous interacting agents. In: Russian Academy of Sciences, Institute of Economics and Central Economics and Mathematics Institute (eds) Economic transformation and evolutionary theory of J. Schumpeter. Institute of Economics and Central Economics and Mathematics Institute of Russian Academy of Sciences, Moscow, pp 265–285

Aruka Y (2004) How to measure social interactions via group selection? A comment: cultural group selection, coevolutionary processes, and large-scale cooperation. J Econ Behav Org 53:41–47

Hare T (1963) Descriptism. Br Acad 49:115–34

Helbing D (1995) Quantitative sociodynamics: stochastic methods and models of social interaction processes. Kluwer Academic, Dordrecht

Hildenbrand W (1983) On the "law of demand". Econometrica 51:997–1008

Hildenbrand W (1994) Market demand. Princeton University Press, Princeton

Hildenbrand W, Kirman AP (1988) Equilibrium analysis: variations on themes by Edgeworth and Walras. North-Holland, Amsterdam

Hoover KD (2001) The methodology of empirical macroeconomics. Cambridge University Press, Cambridge

Mitjushcin LG, Polterovich WM (1978) Criteria for monotonicity of demand functions. Ekonomka i Matematicheskie Metody 14:122–128(in Russian)

Price GR (1972) Extension of covariance selection mathematics. Ann Hum Genet 35:485–490

Riley J (1988) Liberal utilitarianism: social choice theory and J. S. Mill's philosophy. Cambridge University Press, Cambridge

Samuelson PA (1938) A note on the pure theory of consumer's behaviour. Econometrica 5:61–71

Sen AK (1970) Collective choice and social welfare. Holden-Day, San Fransisco

Steindl J (1965) Random processes and the growth of firms: a study of the Pareto law. Griffin, London

Weidlich W (2000) Sociodynamics: a systematic approach to mathematical modeling in the social sciences. Harwood Academic, London

Weidlich W, Haag G (1983) Quantitative sociology. Springer, Heidelberg

Chapter 8
The Moral Science of Heterogeneous Economic Interaction in Face of Complexity

Yuji Aruka

The principles of political economy, particularly those of the Anglo-Saxon origin, were kinds of utilitarian-based moral philosophy. In this sense, the idea of individualistic rational utility is the essential ontological factor of empiricism to generate the source of all human behaviour in the economic system. Thus, many economists have so far been inclined to indulge in an individualistic utility-based prediction, almost everywhere that human nature matters. While the Schumpeterian epistemological view of the continental idealisms is construed as the triangular theoretical layers of statics, dynamics, and sociodynamics. Without any modification of the ontological–epistemological constructs in economics, we can no longer capture the essence of the rapidly salient evolution of complex economic systems in the modern times. Evolutionary economics is a challenge to change the traditional ontological–epistemological set.[1] In this paper, we focus on *heterogeneous economic interaction of agents* to argue *a moral code* for the complex economic system.

[1]As jump-started by Nelson and Winter's book, evolutionary economics promptly gained popularity all over the world at the end of the last century. See Nelson and Winter (1982).

Reprinted from Dynamisches Denken und Handeln. Philosophie und Wissenschaft in einer komplexen Welt. Festschrift für Klaus Mainzer zum 60. Geburtstag. Aruka, Y., The Moral of Heterogeneous Economic Interaction in the Face of Complexity, 171–183 (2007). With kind permission from Hirzel Verlag, Stuttgart.

Y. Aruka
Faculty of Commerce, Chuo University, Higashinakano, Hachioji, Tokyo 192-0393, Japan
e-mail: aruka@tamacc.chuo-u.ac.jp

Y. Aruka (ed.), *Complexities of Production and Interacting Human Behaviour*,
DOI 10.1007/978-3-7908-2618-0_8, © Springer-Verlag Berlin Heidelberg 2011

8.1 The New Features of Preference Theory for the Sequential Choices

8.1.1 Economic Theory in the Complex Economic System

In the last century, the distributive principle of economy has been lost as a result of the rapid structural change of the production process. With "casino capitalism" or the "winner-take -all" society, instead of the classic distributive justice, a lottery system dominates income distribution (Arthur 1996).[2] Even in this situation, orthodox economics still prefers a set of particular rationalities, e.g., the so-called game theoretic views, instead of the general rationality.[3] Rationality, either in general or in a particular form, is not to be regarded as a panacea in the complex socio-economic system.

Mainzer briefly summarized the history of economic theories which, at present, is accepted as the science of rationality.

> Marx erkennt also richtig die nichtlineare Dynamik ökonomischer Systeme, will sie aber durch eine lineare Dynamik im Sinne des laplaceschen Geistes ersetzen. Dazu muss ein neuer Mensch angenommen werden, der nicht nach seinem eigenen Profit strebt, sondern nur die gesellschaftlichen Interessen verwirklichen will. Das ist aber eine unrealistische Annahme der menschlichen Natur. Ebenso geben Adam Smith und die ökonomischen Klassiker von einer idealistischen Annahme über den Menschen aus. Der homo "oeconomicus", der mit vollständiger Information über seine Umwelt nur seinen eigenen Nutzen maximiert und in diesem Sinne nur rational handelt, ist eine mathematische Fiktion linearer Gleichgewichtsdynamik (Mainzer 2005, pp. 60–61).

As Poser (2005) pointed out, advanced research on the complex system by itself will not solve our urgent matters, because human nature always intervenes in the outcomes of society. Thus, research on a new utilitarianism must thus be pursued. In this paper, the use of the utilitarianism of heterogeneous interacting agents is strongly recommended. This new utilitarianism may easily be applied to the transition rates of the master equations, i.e., the probabilistic Markov process. Furthermore, a new method to reconstruct economic science is also suggested: constructing methods derived directly from new ideas in statistical physics and combinatorial stochastic process (Aoki and Yoshikawa 2006). These kinds of attempts could successfully assure the depiction of complex evolving dynamics of the human nature and societies without relying upon the methodological individualism. In this new framework, solidarity formation among the heterogeneous interacting agents should be the most important matter.[4]

---

[2]Arthur depicted this capitalism vividly. Casino capitalism should bring on the "winner-take-all society" if we focus on the distribution side. See Frank and Cook (1995).

[3]These particular rationalities are examined in some detail and their failures are argued in Aruka (2006).

[4]Such a social transition rate as generating solidarity will be operated in the master equation of the social process. According to the steps of the modelling procedure by Weidlich, we will need a set of various personal attitudes in order to set out the social configuration in our model. When we sketch this configuration, we also need the concept of strength and direction of preference, as being

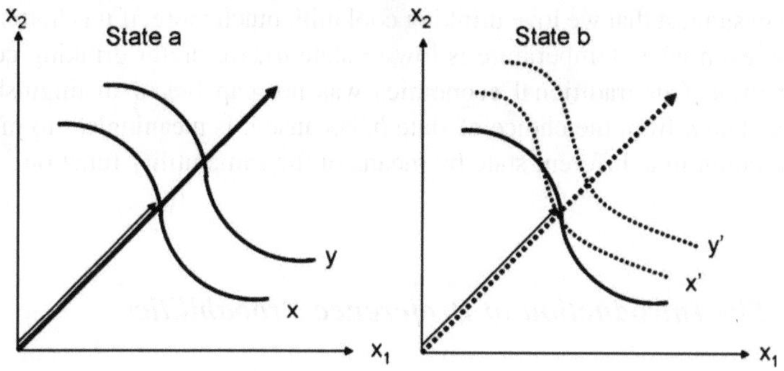

Fig. 8.1 The state-dependent preference

8.1.2 The State-Dependent Preference

We have encountered several difficulties of the so-called traditional preference theory. One objection is a state-dependent preference, as Kreps (1990, Chap. 4) pointed out. A popular utility function is a kind of map from a basket of goods to a specified cardinal number. It is easy to verify that a utility function, even without convex preference, can uniquely give any cardinal number, but only give a number up to scale, i.e., a relative number. An absolute value is meaningless here. This means that any comparison of utility between any two individuals is useless. This similarly leads to making any comparison of utility between *any two different states* useless.

In fact, a series of preference ordering in a state generally is not the same as another series of preference ordering in another state. Given the set of traditional assumptions, a difference of state may characterize a different form of utility function. The range of a utility function in a different state may also be changed. If we had a certain *state transition*, thus, our utility function could not be capable of giving any unique comparison between the absolute values of the concerned two utility functions.

A distance of the arrows in Fig. 8.1 shows a level of utility. It is evident that a different set of indifference curves implies a different utility function. Compare, for instance, the real curves with the dotted curves. Now we suppose that our preference on x, y were unchanged between state a and state b. We then have the formally two similar preference relations but with the two different states:

$$\{x^{-}\succ y |\ \text{state } a\}$$

and

$$\{x'^{-}\succ y' |\ \text{state } b\}.$$

touched later. In particular, direction may be argued as a trend of the social configuration. See Weidlich (2000, Sect. 3.2).

These may suggest that we love drinking cool milk much more, if it is hotter (=state *b*), while, even when temperature is low (=state *a*), we prefer drinking cool milk than hot milk. Our traditional economics was not capable of distinguishing the choice of state *a* from the choice of state *b*, because it is meaningless to measure a similar relation in a different state by means of the same utility function.[5]

8.1.3 The Introduction of Preference Probabilities

8.1.3.1 How to Introduce the Preference Probabilities?

In order to distinguish these states, we may require *the strength of preference* or *probability* to confirm the comprehensive decision making. A weaker preference may correspond to a less probable event. Hence a possibility of *random preference* should be examined. Let the traditional preference relation be as follows:

- ***A-1*** If it is hot, agent A prefers milk.
- ***A-2*** If it is neither hot nor cool, agent A prefers tea.

We may thus formulate the two kinds of constructs for preferences:

- ***B-1*** If it is hot, agent A is more probable to prefer milk.
- ***B-2*** If it is neither hot nor cool, agent A is more likely to prefer tea.
- ***C-1*** If it is hot, many prefer milk or few prefer tea.
- ***C-2*** If it is neither hot nor cool, many prefer tea or few prefer milk.

Random preference, at first glance, indicates the cases *B*-1 and *B*-2. We may, however, have another interpretation by *framing*. In the next section, we will argue the cases *C*-1 and *C*-2. Noticed that the idea of *subjective* agent is indispensable in the cases *A* and *B*. Here a subjective agent implies one who aims at *optimizing* his/her subjective object. Alternatively, we do not necessarily need such an agent in the case *C*. In the latter setting, we may argue *the state variables* like *type distribution* of agents *in transition*. This argument leads to giving up *the individualistic subjectivism* (Fig. 8.2).

[5] In addition to *such a state transition*, we will use the information on many attributes of the individual like age, gender, income to refine an actual utility function: "The analyst has to identify those that are likely to explain the choice of the individual. There is *no automatic process* to perform this identification. The knowledge of the actual application and the data availability plays an important role in this process". Bierlaire (1997, p.4).

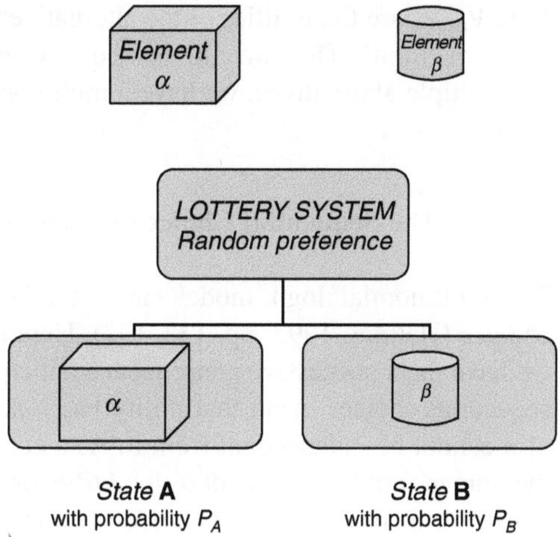

Fig. 8.2 Random preference

8.1.3.2 Bohr Versus Einstein?

As for the source of *uncertainty*, according to Anderson et al. (1992, pp. 19–20), we have two different kinds of ontology: the first one is that *decision rule* are *intrinsically stochastic*, while the utility is stochastic, as argued by Luce (1959) and Tversky (1972), the other is that the decision rule is deterministic, while the utility is stochastic, as when generating the random utility model or the neoclassical model. The difference between both models is just analogous to the debate between *Bohr* and *Einstein* about *the uncertainty principle* (Bierlaire 1997, p. 8).

8.1.4 The Sequential Choice Formulation

8.1.4.1 The Multi nomial Logit Model

As already known in the field of discrete choice pioneered by Luce, a psychologist, *the multinomial logit model of utility* can, *remarkably*, be derived *both* from *the random utility theory* and from *the discrete choice axiom.*

We have the model *the binary model*, when we only have binary alternatives $\{a, b\}$. In the Luce model, the probability of an alternative a is expressed by the equation:

$$P_{\{a,b\}}(a) = \frac{1}{1 + e^{-(\mu V_a - \mu V_b)\mu}} = \frac{e^{\mu V_a}}{e^{\mu V_a} + e^{\mu V_b}}. \tag{8.1}$$

Here V_a, V_b are the utilities of the alternatives a, b, respectively. The equation is of the form "logit". Thus we call the above form (8.1) "a binary logit model". If there are multiple alternatives, we have a multinomial logit model.

8.1.4.2 The Sequential Choice in Terms of the Multinomial Logit Model

The multinomial logit model can actually implement the world of sequential choices (Helbing 1995, pp. 132–137). Here *the decision* for agent i is *done* when we have the k successive consequence anticipations on the same alternative x. The sequential choices mean that utility has *a past history of decision sequences* and also cannot be independent from it. As a consequence, utility will *fluctuate* like in the human mind. We may describe *utility* or *preference* of decision on the alternative x for agent i whose last choice was the decision y by

$$U^i(x|\,y;t).$$

The probability for occurrence of the imagined consequence$\imath$, due to $f(\imath) = x$, will be

$$p^i(x|\,y;t).$$

We then apply *the utility of multinomial logit model* to the probability:

$$p^i(x|\,y;t) := \frac{\mathrm{e}^{U^i(x|\,y;t)}}{\sum_{x'} \mathrm{e}^{U^i(x'|\,y;t)}}. \tag{8.2}$$

Hence the probability that the next k arguments are all in favour of the alternative x will be as follows[6,7]:

$$p'(x|\,y;t) = [\hat{p}(x|\,y;t)]^k = e^{k\hat{U}(x|\,y;t)}. \tag{8.3}$$

We define the *attractiveness* of the decision x for agent i who chose the behaviour y during k periods as follows:

$$U'(x|\,y;t) := k\,\hat{U}(x|\,y;t) - \ln \sum_{x'} e^{k\,\hat{U}(x'|\,y;t)}. \tag{8.4}$$

Now we may create a new idea, like a factor expressing some psychological trend, according to Helbing (1995). He actually defined the *readiness on decision* x':

[6]Here i is omitted for simplicity.

[7]We use the scaled preferences such as $\hat{U}(x|\,y;t) := U(x|\,y;t) - \ln \sum_{x'} e^{U(x'|\,y;t)}$.

$$R(x'|x;t) := \frac{e^{U(x',t)-U(x,t)}}{D(x'|x;t)}. \tag{8.5}$$

Here D is the distance between x and x' at time t, and it must be

$$D(x',x;t) = D(x,x';t) > 0.$$

8.1.4.3 The Social Transition Rate of the Subgroup in the Community

So far we discussed the above arguments at the level of the individualistic decision of agent between x and x'. As we learn precisely in the next section, the so-called universality of the individualistic agent, when s/he is faced with the changing environment exposed by heterogeneous interaction between agents, does not always hold at any time. Thus we may aggregate agent i into a subgroup σ. We may then express *the social transition rate* for a subgroup σ, w_σ from x to x' in terms of the new factors of *flexibility* v and *readiness* R, for example:

$$w^\sigma(x'|x;t) = v(t)R^\sigma(x'|x;t) \tag{8.6}$$

for $x' \neq x$. This formulation suggests a benchmark to depart from the individualistic subjectivism.

If our solution were inclined to be fixed in the long-run solutions, v and R^σ could contribute to *the trend* of preference of the subgroup σ; *attractiveness* could make out *the strength* of preference for the subgroup σ. These ideas may directly contribute to the modelling of *spatial and urban development* in particular.[8]

8.2 What Is Heterogeneous Interaction?

8.2.1 How to Formulate the State of Affairs?

Before the case C *in the above section*, i.e., a so-called stochastic process for this argument, is discussed, the two important ideas – of the exchangeability of agents and of the multiplicity of microeconomic configurations – must be understood.

[8] See Sects. 9.1–9.4 in Weidlich (2000).

8.2.1.1 Exchangeability Between Agent *i* and *j*

In a multi-agent system, in general, we cannot immediately determine which agent will achieve the best performance (Aruka 2003). This implies that an agent of a *type* i will be the agent of another *type* j in course of time, e.g., via *imitation*. Here i can be exchanged with j. The idea of *a master equation* depends on this kind of exchangeability.[9] It thus is noted that exchangeability may impose a double meaning on a state.[10]

8.2.1.2 Multiplicity of Microeconomic Configurations

The total number of states is set N. The $n(t)$ indicates the state variable. There are $\frac{N!}{n!(N-n)!}$ ways to realize the same n/N. In other words, the equilibrium distribution gives the idea of a multiplicity of micro-configurations that produces the same macro value. We will not discuss more due to space constraints.

8.2.1.3 Exchangeable Agents in the Combinatorial Stochastic Process

The exchangeable agents emerge by the use of *a random partition vector*, as in statistical physics or population genetics. The partition vector provides us with *the state information.* Thus, the size-distribution of the components and their cluster dynamics can be argued. Define N as its most countable set, in which the probability density of transition from state i to state j is given, respectively. In this setting, the dynamics of the heterogeneous interacting agents gives *the field* where an agent can become another agent. It is also important to note that this way of thinking easily welcomes the *unknown agents*. For example, in the financial markets, there are two different firms: the firm self-financed for investment, and the firm indirectly financed. Focusing on firms of the same size capitalization for the two types, in this partition, we could track the case of bankruptcy of the latter type firm as the result of bank loan diminution in the birth–death process of stochastic dynamics.

---

[9]We define a Markov chain X_t on the state space S which dynamically describes a flux of probabilities. In a state j at time t, we can imagine a probability $P_j(t) = \Pr(X_t = j)$. We suppose that there are a number of independent agents, each of whom is in one of finite microeconomic states, and its state evolves according to the master equation

$$\frac{dP_j(t)}{dt} = \sum_{k \neq j} [P_k(t)w_{kj} - w_{jk}P_j(t)] \text{ for j} \in \text{S},$$

where w_{kj} denotes the transition rates from state j to state k, or the in-flow rates, and w_{jk} the rates from k to j, the out-flow rates.

[10]The applications of this method are strenuously developed by Aoki (2002), for instance.

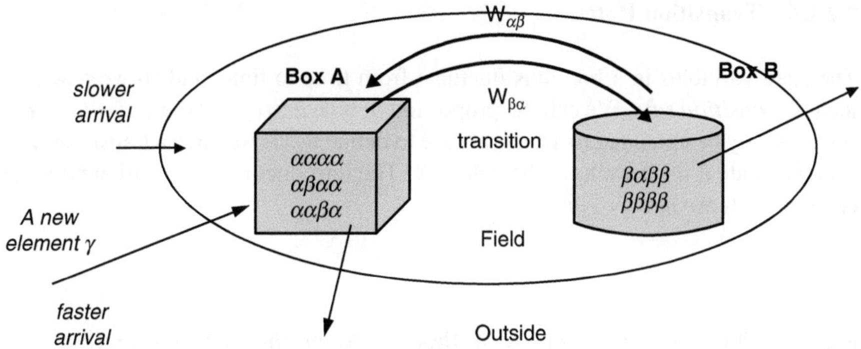

Fig. 8.3 An idea of the master equation

8.2.1.4 Types in Transition

In a simple setting of random preference as illustrated in Fig. 8.3, it only matters for the concerned agent whether either α or β happens as a consequence of randomizing elements $\{\alpha, \beta\}$ like in a lottery system. Now suppose that an agent priority or type cannot be determined without its particular field as well as its state. Taking into account *different types in transition* may lead to bringing us another modelling, typically represented by a master equation. Type α in a box A may move to another box. Such an *exchangeable type* may contribute to form *a state property.*

8.2.1.5 Heterogeneous Interaction as a Type Distribution

We depict our setting in the following manner: We have a box to accept various types. In other words, each type can move into any box. So we have each type distribution in each box. Types in a box are regarded as *the state variables*. We also consider a kind of *heterogeneous* element to be a type distribution. In the case of C as mentioned above, each element or *type* $\{\alpha, \beta, ...\}$ in a box may be interpreted as the first priority of each agent. Given *the transitions of types between boxes,* however, the majority of type α in box A may collapse due to the immigration of a heterogeneous type β into box A from box B. Similarly, the state will be affected by due to emigration from box B. These parts reflect *heterogeneous interaction.* In Fig. 8.3, box A shows a stable state property, if type α dominates in box A; box A shows *a critical state* property, if we have α to $\beta = 0.5$ to 0.5 in box A, because box A can no longer be discriminated against box B. If type α dominates, box A and state A cannot be distinguished.[11]

[11]If we had this critical situation, we would have a kind of entropy maximization.

8.2.1.6 Transition Rates

The state variables in a box thus fluctuate from time to time, and so we reach the idea of *transition rate*. We call the proportion of type α to type β the *transition rate*. It may easily be extended to a general case to define w_{ij}, if we have a transition from state j to state i; w_{ij} if we have the converse. The introduction of transition rates can channel with *the master equation*.

8.2.1.7 The Comparison of the Methods Between the Master Equation approach and Subjectivism

Finally we can summarize about the comparison between *the master equation approach* and a traditional *individualistic subjectivism* in Fig. 8.4. In the latter, by the assumption of uniform similarity among agents, i.e., homogeneity among agents, *a representative agent* works well for his/her individualistic optimization in all the fields of the economic system.

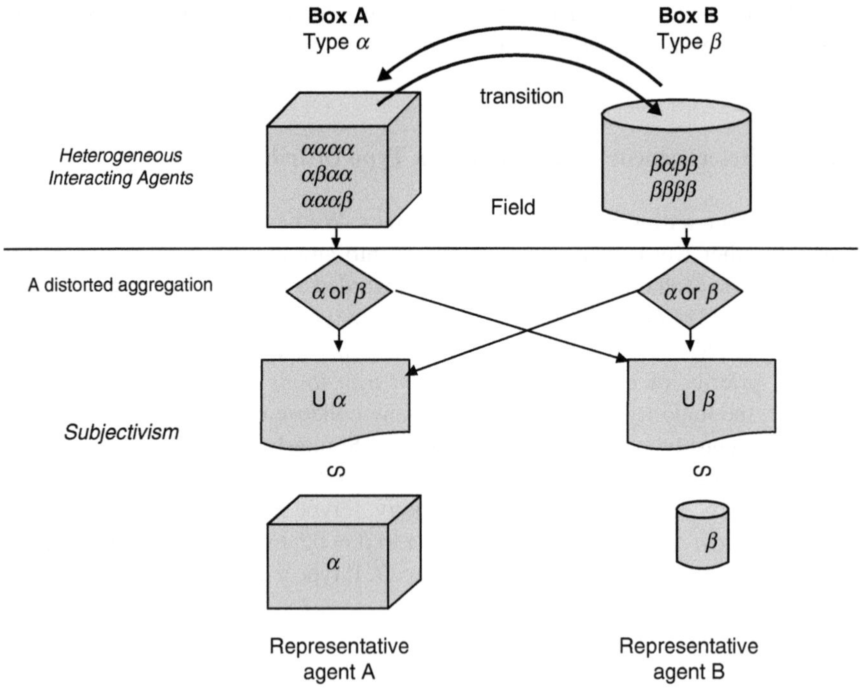

Fig. 8.4 The comparison of the methods

8.3 Altruism and Coordination in the Community of Heterogeneous Interaction

8.3.1 *Direction of Preference*

Utility maximization by itself is not a presentation of selfishness, because any device to measure how much his/or her maximization affects others' utility is not installed at all in our individualistic utility function. In order to argue selfishness, our utility must be modified. A solution for this problem is to argue the strength and direction of preference.

Selfishness cannot be defined without the interactions between the agents.[12] The direction of strategy will depend on the situation of community as a whole. In fact, in the *Avatamsaka game* (Aruka 2001), my original game, selfishness would not be determined even if the agent selfishly adopted the strategy of *defection*. Individualistic selfishness could only be realized if the other agent *cooperates*. Any certain gain from defection can never be assured by defection alone. *The sanction by defection* as a reaction of the rival agent does never imply the selfishness of the rival (Fig. 8.5).[13]

The strength and direction of preference may thus be indispensable to formulate the evolution of utility. Preference may therefore be accompanied by *the trend* of selfishness. Selfishness may be a basic trend of the human mind (Fig. 8.6).

Strategies	Player B	
Player A	Defection	Cooperation
Defection	(0, 0)	(1, 0)
Cooperation	(0, 1)	(1, 1)

Fig. 8.5 Avatamsaka game

[12]The rational postulates on individualistic behaviour will not necessarily guarantee selfishness of the result of choice. Bowles and Gintis ask in the following manner: "Is cooperation, then, just an expression of self-interest? Do other-regarding preferences such as a reciprocity and altruism play no role in the explanation of human cooperation? The answer hinges on whether most forms of cooperation are altruistic [in society] or mutualistic [in biology]?" (Bowles and Gintis 2005, p. 22).

[13]It is noted that Akiyama smartly defined the class of games as dependent games such as Avatamsaka games. See Akiyama and Aruka (2006).

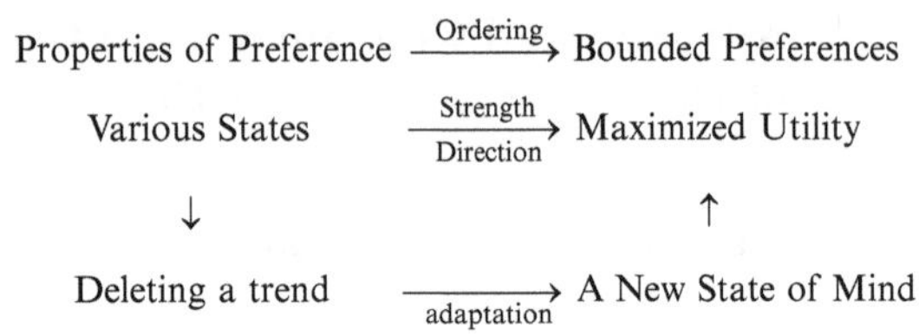

Fig. 8.6 A new scheme in the utility function

8.3.2 The Mozi: Impartial Caring for Heterogeneous Interacting Agents

8.3.2.1 Mozi

Two heterogeneous interacting agents will be faced with a branching stochastic process where a usual individualistic optimization fails. Given such a branching point, the so-called utilitarianism of homogeneous producers and consumers cannot help breaking down. We thus are keen to construct the utilitarianism of heterogeneous interacting agents. We actually trace this idea to its ancient Chinese origin: the Mozi School; Mohism, appeared in about 500 BC. Mozi (ca. 470 BCE until ca. 390 BCE)[14] declared that a big power should not attack a small country. He simultaneously used enough full of his highly professional knowledge and skills as a munitions or fortifications expert so that he could practically force the generals who attempted to invade a small state into giving up merely by the use of war game on the desk. This school was also well known as anti-Confucianism. The idea of impartial caring is often appreciated as the same idea as the Western utilitarianism. However we must notice that the Mozi utilitarianism was not based on the individualistic homogeneous agents. Mozi rather had found a definite idea of multiple stake holders of the modern corporation system in terms of the ancient heterogeneous agents, and also added to an original idea of a moral belief for coordination.

> The Mozi school did not focus on the so-called naked individuals when it developed the idea of impartial caring. It regarded the world as the aggregate of seven different intersecting units (father–sons, elder brother–younger, lord–subjects, myself–others, wife–other wives, family–other families; home country-foreign countries): within each unit, unit members share the same interest, and between units, they stand in opposition. ... The Mozi school classified the world in light of the only criteria to maintain social order, according to its original purpose of impartial caring. ... It asserted that there was no discrimination between one's own unit and an opponent's unit (impartial); any agent should not attain his own benefit at the price of his opponent (caring or regarding others) (Asano 1998, p. 59).

[14]See the full text of Mozi (Chinese and English) at http://chinese.dsturgeon.net/text.pl?node=101&if=en

8.3.2.2 The System of Mutual Benefiting as a Moral Code

In the earlier form of the Mozi school, Mozi's idea of rejecting the envy of profit simply implied the negation of private profit. In a more advanced form, the school advocated *the system of mutual benefit*. In this stage, the profit is strictly related to others' profits.

The system of mutual benefit aims at a situation that all the people could enjoy the whole welfare of a world just recovered when the activities of pursuing self-interests at the expense of others entirely ceased, according to *the spirit of impartial caring* (Asano 1998, p. 59).

Thus we see that the constituents for modelling the heterogeneous interacting agents are implemented in the Mozi school as discussed in the section above:

- Community members functionally are classified into heterogeneous agents.
- Heterogeneous agents aggregate into subgroups as clusters, respectively: agent belongs to a cluster in which unit members' interest is entirely *unanimous*, while each cluster conflicts with each other.
- The profit may be exchangeable between clusters: a system of mutual benefiting. This suggests an idea of exchangeable agent as previously argued.

Proposition. *The framework of heterogeneous interacting agents can employ the system of impartial caring and mutual benefit as its moral code.*

The transition rates of the master equation implies the in-flow rate and the out-flow rate, describing by themselves *the forces of attracting and repelling*. As we have seen, these rates can be implemented by the multinomial logit utilities. These renewed rates – equipped with the attractiveness, readiness and so on – can thus contribute to depicting the successive process of social decisions reflecting the different kinds of heterogeneous interaction. This is just the economic implication of *sociodynamics* in terms of moral science, and this type of analysis has already been developed by Weidlich and Haag working on *the synergetic project* in Stuttgart. Thus we can argue the solidarity formation in connection with an actual historic path. In other words, sociodynamics in terms of the Mozi system of thought implies the feasibility of numerical analysis, instead of the system of representative agent equipped with the invisible hands.

References

Akiyama E, Aruka Y (2006) Evolution of reciprocal cooperation in the Avatamsaka game. In: Namatame A, Kaizoji T, Aruka Y (eds) The complex networks of economic interactions: essays in agent-based economics and econophysics. Springer, Heidelberg, pp 307–320

Anderson SP, de Palma A, Thisse J-F (1992) Discrete choice theory of product differentiation. MIT, Cambridge, MA

Aoki M (2002) Modeling aggregate behavior and fluctuations in economics: stochastic views of interacting agents. Cambridge University Press, New York

Aoki M, Yoshikawa H (2006) Reconstructing macroeconomics: a perspective from statistical physics and combinatorial stochastic processes. Cambridge University Press, Cambridge, NY
Arthur WB (1996) Increasing returns and the new world of business. Harvard Bus Rev 74:100–109
Aruka Y (2001) Avatamsaka game structure and experiment on the Web. In: Aruka Y (ed) Evolutionary controversies in economics. Springer, Tokyo, pp 115–132
Aruka Y (2003) The complex adaptive processes in economics by heterogeneous interacting agents. In: Russian Academy of Sciences: Institute of Economics and Central Economics and Mathematics Institute (ed) Economic transformation and evolutionary theory of J. Schumpeter. Institute of economics RAS, Moscow, pp 265–285
Aruka Y (2006) Evolution der Sittenlehre über Wirtschaftliche Rationalität im Komplexen Sozialsystem. Symposium zur Gründung einer Deutsch-Japanischen Akadmie für integrative Wissenschaft
Asano Y (1998) Mozi. Kodansha, Tokyo (in Japanese)
Bierlaire M (1997) Discrete choice models, (mimeo) (http://web.mit.edu/mbi/www/michel.html)
Bowles S, Gintis H (2005) Can self-interest explain cooperation? Evol Inst Econ Rev 2(1):21–41
Frank R, Cook P (1995) The winner-take-all society. Free, New York
Helbing D (1995) Quantitative sociodynamics: stochastic methods and models of social interaction processes. Kluwer Academic, Dordrecht
Kreps DM (1990) A course in microeconomic theory. Princeton University Press, New Jersey
Luce R (1959) Individual choice behavior: a theoretical analysis. Wiley, New York
Mainzer K (2005) Was sind komplexe Systeme? In: Symposium zur Gründung einer Deutsch-Japanischen Akadmie für integrative Wissenschaft, J.H. Röll Verlag, pp 37–77
Nelson RR, Winter SG (1982) An evolutionary theory of economic change. Belknap Press of Harvard University Press, Cambridge, MA
Poser H (2005) The prediction problems in the complex sciences, in complexity and integrative science. Koyo-Shobo, Kyoto, pp 3–26 (in Japanese)
Tversky A (1972) Elimination by aspects: a theory of choice. Psychol Rev 79:281–299
Weidlich W (2006) Sociodynamics: a systematic approach to mathematical modeling in the social sciences. Harwood Academic Publishers, Amsterdam (The Gordon and Breach Publishing Group). [Reprinted by Taylor and Francis (2002); Paper edition, Dover Publications (2006); Japanese translation, Morikita Shuppan (2007)]

Chapter 9
The Evolution of Moral Science: Economic Rationality in the Complex Social System

Yuji Aruka

Abstract Economics in the early 20th Century established distributive justice as the marginal productivity theory of income distribution. As the system has evolved, however, the distributive principle has been lost as a result of the structural change of the production process. Faced with "casino capitalism" or the "winner-take-all" society, instead of the classic distributive justice, a lottery system dominates income distribution. Orthodox economics prefers a set of particular rationalities, e.g., the so-called game theoretic views, instead of the general rationality. These particular rationalities are examined in some detail and their failures are argued. Rationality, either in general or in a particular form, is not to be regarded as a panacea in the complex socio-economic system.

This paper proposes the use of the utilitarianism of heterogeneous interacting agents. This new utilitarianism may easily be applied to the transition rates of the master equations, i.e., the probabilistic Markov process. Furthermore, a new method to reconstruct economic science is also suggested: constructing methods derived directly from new ideas in statistical physics and combinatorial stochastic process.

In sum, individualistic rationality must be replaced with the utilitarianism of heterogeneous interacting agents. In this new framework, solidarity formation among the heterogeneous interacting agents should be the most important matter. Finally, a deeper consideration on the utilitarianism of heterogeneous agents is explored.

This paper is mainly based on the paper "Evolution der Sittenlehre über Wirtschaftliche Rationalität im Komplexen Sozialsystem" presented at 3. Wissenschaftliches Symposium, Deutsch-Japanische Gesselschaft für integrative Wissenschaft, Montag, 30. Oktober 2006, Museum Koenig, Bonn, though this is not the same in is contents. The author is very grateful for Abt Nissho Takeuchi, the German–Japan Society for Integrative Science, Bonn, and the Daiseion-ji, Wipperfürth for generous permission of this material.

Reprinted from Evolutionary and Institutional Economics Review 4(2), Aruka, Y., The Evolution of Moral Science: Economic Rationality in the Complex Social System, 217–237 (2009). With kind permission from Japan Association for Evolutionary Economics, Tokyo.

Y. Aruka
Faculty of Commerce, Chuo University, Higashinakano, Hachioji, Tokyo 192–0393, Japan
e-mail: aruka@tamacc.chuo-u.ac.jp

Y. Aruka (ed.), *Complexities of Production and Interacting Human Behaviour*,
DOI 10.1007/978-3-7908-2618-0_9, © Springer-Verlag Berlin Heidelberg 2011

Keywords Failures of individualistic rationality · Heterogeneous interaction · Moral science · Sociodynamics · Statistical physics

9.1 Introduction

Our society is a huge system, more or less, like the universe. We, however, must not encounter the "heat death" of society. Otherwise, in the case of the human community, we are faced with dissipative self-organizing where we would usually observe the forming of a certain order of human organizations, accompanied by various series of fluctuations. This paper looks at modeling the idea of sub-group dynamics in the socio-economic system. Given such a dissipated system in the society and the economy, the variously biased distributions of the system, irrespective of the aggregation of individually optimizing behavior, would often be observed.

The increase in the number of empirical studies on the price fluctuations of securities, equities, and land, as well as the size distributions of capital and sales, have systematically been observed by several groups of physicists. This group activity is called "econophysics" – a new economic science classified as heterodox economics by the standards of the American Economic Association. In short, the new economic sciences are evolving by integrating several related interdisciplinary sciences.

As Poser pointed out, however, advanced research on the complex system by itself will not solve our urgent matters (Poser 2005). Human nature always intervenes in the outcomes of society. Economics is itself a science of human nature, i.e., a moral science, necessary in the complex social system. Mainzer (2005) briefly summarized the history of economics which, at present, is accepted as the science of rationality. In this paper, the limitations of rationality are discussed.

9.2 The Neo-classical Principles of Political Economy

Erich Schneider, a distinguished German economist, uniquely constructed a modern form of microeconomics in the 1930s (Schneider 1934). In the first half of the last century, however, "economics" in the English speaking culture was sharing the double name of economics and political economy. The University of Chicago, for example, is still proud of publishing a distinguished journal, *The Journal of Political Economy*.

9.2.1 The Individualistic Principle

In the beginning of modern or neo-classical economics, departing from the classical principles of political economy, pioneering economists declared the idea of distributive justice in accordance with the theory of Vilfredo Pareto, which implies that the improvement of personal income distribution or welfare must not impose upon every level of any individual.[1] These economists did not think of the market as the sole means to realize the distribution of income in the economy.

There is actually another route to improve personal welfare. Even if the market mechanism fails to guarantee the improvement of welfare for everybody, these pioneers could have alternatively suggested improvement by a coalition of agents, i.e., negotiation, in the specified field for trading where any redistribution of the initial welfare can bring everybody more welfare than before. The coalition solution, within the limits of redistribution, is capable of attaining the same achievement as the market mechanism, provided that the economy is just a "replica economy" where numerous twins actually live, as the limit theorems on *The Core Economy* indicate.[2] However, it is difficult to find the realistic conditions in which the market solution coincides with the coalition solution. If certain heterogeneous agents joined in, the conditions could be easily blocked.

A principle of individualism was first stated by the utilitarian propositions that mean "subjective egalitarianism", i.e., every agent can maximize his/her utility subject to his/her personal budget. Any consumer in any time, if he/she were intelligent, would be permitted to realize his/her individual optimization, given the market prices common to every individual. He/she can always be treated as completely equal, to the degree that he/she has the same ability as well as the same initial wealth. The great success of economic rationality has been to discover the "individualistic utility function" whose properties are continuous and C^n-differentiable, and whose derivatives, traditionally, are monotonically decreasing its arguments.

9.2.2 The Marginal Productivity Theory

Even at this stage on the consumption side, there could not be full-fledged distributive justice for income. In order to reach this, the utility function must be used to formulate the production function, because the real income of a firm is created by production, and the personal income by an individual's labor. In particular, total cost minimization, subject to the set of admissible techniques in the form of a production function, may be attained. Here, the environments of productive technologies should be accepted where the envelop theorem in the functional analysis could hold. The long-run true minimal total cost to establish the optimal level of production can then be found, provided that firms can compete perfectly

[1]The Pareto criterion originally implied that the personal welfare of every agent must not be impaired absolutely.

[2]*Core* is defined as a Pareto optimum within the area of being at least no worse than the initial welfare level.

Fig. 9.1 A single productive process

$$\begin{pmatrix} x_1 \\ \vdots \\ x_i + \delta x_i \\ \vdots \\ x_m \end{pmatrix} \Rightarrow y + \delta y \text{ due to } \delta x_i$$

given the price of the product common to all the numerous participating firms. Correspondingly, in the "market of factors" like capital or labor, each firm, to achieve the optimal production, can set the prices of its factor inputs to its marginal productivities, given the factor prices common to all the participating firms.

In the classical single production function characterized by neo-classical, "one-way street" production where inputs are the factors of production, never entering the outputs, the process i is as follows:

$$Input(x_1, ..., x_i, ..., x_m) \Rightarrow Output(0, ..., 0, y_i, 0, ..., 0) \tag{9.1}$$

Other things being equal, the marginal variation of the output can uniquely be assessed by an additional particular factor input $x(i)$. As far as this holds, the distributive justice of income is assured by the marginal productivity principle.

In this world, every additional personal effort can be afforded its remuneration any time in real terms. Only a man who spares his own trouble cannot earn his reward. *If you won't work you shan't eat* (Fig. 9.1).

9.3 Market Capitalism Without Distributive Justice

As the system evolves, a simple agent of the system becomes adaptable to a more hierarchical configuration. In due time, a homogeneous agent may become a heterogeneous interacting agent working in a more extended field. A complex economic system may be characterized by a set of sub-groups of different types. If this evolution could be applied to the market, the ways of competition would be changed. At the end of the previous century, the so-called "casino capitalism"[3] or, equivalently, the "winner-take-all society"[4] appeared. This capitalism holds, irrespective of Pareto improvements among the participating agents.

[3]Arthur (1996) depicted this capitalism vividly.

[4]Casino capitalism should bring on the "winner-take-all society" if we focus on the distribution side. See Frank and Cook (1995).

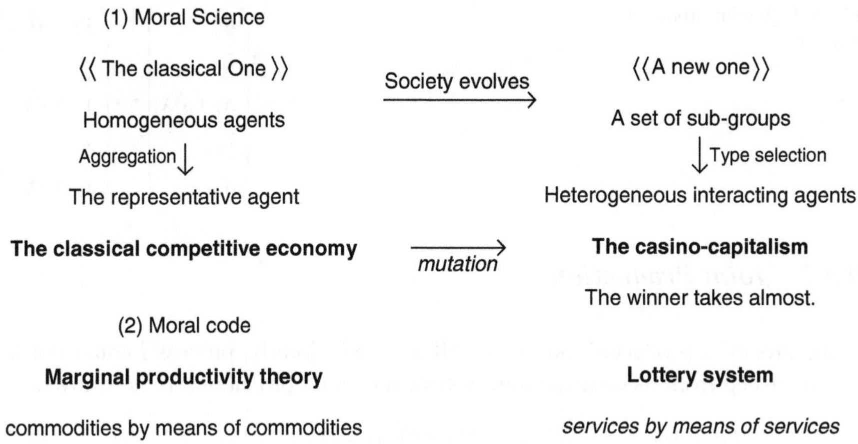

Fig. 9.2 A scheme of moral science evolution

9.3.1 The Collapse of the Neo-classical Distributive Principle

The collapse of the neo-classical distributive principle takes place in a more general production function, especially in the case of "the production of services by means of services". This may eventually lead to the adoption of the lottery system of distribution.[5]

These situations are depicted in Fig. 9.2.

The reason why the Pareto improvement no longer holds may *a fortiori* be reinforced by the special feature of production, which is signified by the "production of services by means of services" where the marginal productivity principle fails. In modern times, we are used to being involved in a kind of "joint-production". Production, in general, shares almost the same process observed in a chemical reaction. The oil processing industry represents a classic example. A contemporary example is the information industry, incessantly reproducing by networking. Here the relationship between causes and results may be quite ambiguous.

Remark 1. In the last century, we have experienced the rapid change of industry composition from a manufacturing to a service industry. In the advanced countries, it is common that the proportion of the economy that is service industry outweighs that of all other industries. Economics, however, lacks the theory of service production by means of services. In traditional economics, services are created by factors of production like labor, land, and capital. On the contrary, our consumption now is composed of services. It thus is noteworthy to notice that IBM's Almaden Research Center, in November 2004, proposed a new science called service science.

[5]In the classic case, we called the irreproducible factors like labor or land the factors of production. In modern services production, the concept of labor may evolve.

Fig. 9.3 A joint-productive process

$$\begin{pmatrix} x_1 \\ \vdots \\ x_i + \delta x_i \\ \vdots \\ x_m \end{pmatrix} \Rightarrow \begin{pmatrix} y_1 + \delta y_1 \\ \vdots \\ y_i + \delta y_i \\ \vdots \\ y_n + \delta y_n \end{pmatrix}$$

9.3.2 *Joint Production*

If the process i produced good i as well as good j jointly, process j could not be distinguished from its contribution. A system of joint production is as follows:

$$Input(x_1, ..., x_i, ..., x_m) \Rightarrow Output(y_1, ..., y_{i-1}, y_i, 0, ..., y_{n-1}, 0) \tag{9.2}$$

In such a circumstance as in the above general productive process, there is not only the impossibility of neo-classical distributive justice but there is also a series of various perverse phenomena like negative outputs, as well as negative prices (Fig. 9.3).[6]

9.4 Particular Economic Rationalities and Their Failures

We have seen that the core economy via coalition formation does not always coincide with the market solution. This means that the market was separated from the welfare improvement of the Pareto principle. Hence the so-called "invisible hand" no longer holds, generally. The market only functions as a lottery system. Moreover, it is important to know that there cannot be a general principle of economic rationality based on distributive justice. The modern mode of production does not support the marginal productivity principle. Orthodox economics, however, still clings to the macroscopic principle of marginal productivity. Regrettably, the complex economic system will prevent the principle in its aggregate form from being active.

In conclusion, on the one hand, the concepts of utility and production functions are too limited to generalize as well as to adapt to the rapid structural changes of economic systems. Rather, our limited device is expected to change into a new, more promising one. On the other hand, a particular principle of economic rationality must be found in each particular case. This means that there may be some multiple solutions on rationality, leading to some conflicting ends. In the next sections, some well-known economic rationalities are briefly discussed: the Expected Utility Principle, the Full Disclosure Principle, Lemon's Principle, the Contested Garment Principle, and the Loss Minimization Principle.

[6]As for the perversities of the joint production, see Schefold (1989). Many orthodox professional economic theorists have made efforts to generalize the mathematical conditions to contain the non-convexity of production and consumption, without paying attention to the possibility of the distributive justice of production.

Table 9.1 The two choice profiles containing a dangerous event

Profile	▲$1 million	$1 million	$10 million
Profile A	0.01	0.99	
Profile B	0.1		0.90
	Present danger	Social alternative A	Social alternative B
Future scope		Availability on A	Availability on B

9.4.1 *The Expected Utility Principle*

This principle is popular because of its convenience. However, it contains a certain fatal failure pointed out by Allais' paradox It also fails in another way.

When an agent is facing great damages, bankruptcy, or the possibility of death – in other words, when an agent is confronted with real danger instead of manageable risks – the expected utility calculation may be useless to judge the two profiles. It may happen that people will never prefer Profile B to Profile A, even if the expected utility of Profile B is much greater than Profile A's (Table 9.1).

Proposition 1. *If any macroscopic order influences the trend of the system, the expected utility may be useless.*

9.4.2 *The Full Disclosure Principle*

> Individuals must disclose even unfavorable qualities about themselves, lest their silence be taken to mean that they have something even worse to hide. (Frank 2003, p. 201)

The full-disclosure principle derives from the fact that potential adversaries do not all have access to the same information. In the toad case, the asymmetry is that the silent toad knows exactly how big he is, while his rival can make only an informed guess.[7]

In this case, it is very important to know that a threshold is structurally stable given a disturbance to the system. However, a threshold in a nonlinear system is quite vulnerable. Thus, the next proposition:

Proposition 2. *In the self-organizing system, a series of fluctuations may invalidate the supposed thresholds. A supposed threshold may not be so precisely predicted so as not to prevent agents from disguising themselves.*

[7]This principle is active in a choice under uncertainty. We take an example of toads croaking. We suppose that the threshold pitch of croaking is set at 6.0, the lowest at 0, the highest at 10. "If only those toads with a pitch below 6.0 bother to croak, toads who remain silent reveal that their pitch is, on the average, significantly higher than 6.0"(Frank 2003, p. 202).

9.4.3 The Lemon's Principle

Generally speaking, the so-called informed guess of a particular agent could often not hold, if the stage in which a particular rationality is applied were shifted a little by some disturbance.

Asymmetry often causes inconvenience in our secular world, especially the asymmetry of information among participating agents. The well-known "Lemon's Principle" may be classified as this kind of principle. It suggests that it is rational for an owner of a good car to leave the used car market, because a layman cannot distinguish a better used car from a "lemon" (i.e., a bad used car).

It may however be expected that:

Proposition 3. *As the expert system evolves, the market can be graded, creating a segmented market for a good car. The rationale for the owner of a good used car, based on Lemon's Principle, might cause him to lose his sales opportunity.*

9.4.4 The Contested Garment Principle

Behavioral economists often criticize rationalities. Among those economists, the ultimatum game is quite well-known. This game deals with the rationality of 1¢.

> Suppose that you are given €100. You have to offer a part of this amount to me, while I have the right to veto your offer. Given a veto, you will lose your money. You have one chance to make an offer. How much should you offer to me? The solution in view of economic rationality is to offer just 1¢ to me. This is not a realistic solution in this secular world, as many experiments have shown. A person sometimes even offers €60.

The rationality that an agent can demand his maximal share by offering 1¢ to the other agent is called the Contested Garment Principle when the allocation is argued in the Mishnah. The Mishnah is the core text of the Talmud, which is a record of rabbinic discussions of Jewish law, ethics, customs, and stories that are authoritative in Jewish tradition.

Suppose that the man whose debt exceeds his property dies. Here are three creditors *a,b,c* whose claims are d_a, d_b, d_c (Table 9.2).

The principles appearing in the table seem different. People had actually failed to grasp the true meaning of these differences using the principles above for a fairly

Table 9.2 The allocation of the assets in the Mishnah

Creditor	*a*	*b*	*c*	Allocation principle
Total Assets *E*	$d_a = 100$	$d_b = 300$	$d_c = 500$	
100	$\frac{100}{3}$	$\frac{100}{3}$	$\frac{100}{3}$	Equal
300	50	125	125	*Not clear*
450	50	150	250	Proportional

long time. Surprisingly, however, it turns out that all these allocations satisfy the common properties. This was proven by Aumann and Maschler (1985).

9.4.4.1 The Collection Allocation as Well as the Shortage Allocation According to the Size Ordering

Let x_i be the collection [= receipt] of creditor i. Let the debt set $\{d_a, d_b, d_c\}$ as follows:

$$0 \leq d_a \leq d_b \leq d_c.$$

It is easy to see that it holds

$$0 \leq x_a \leq x_b \leq x_c, \tag{9.3}$$

$$0 \leq d_a - x_a \leq d_b - x_b \leq d_c - x_c. \tag{9.4}$$

Each creditor is allocated according to the size ordering of the claims. The shortage of each creditor also is allocated according to the size of the claims. That is to say, the shortage of collection for the creditor who has a bigger claim will be bigger. This represents monotonic increasing of collection x_i in respect to E. As the total assets E (i.e.,$=\Sigma_i x_i$) increases, the collection of creditor x_i must not be decreased.

$$\frac{\partial E}{\partial x_i} \geq 0. \tag{9.5}$$

Lemma 1. *If each player is given his guaranteed minimum, the remaining surplus of the total assets after deleting their guaranteed minima can be shared by a coalition formation among players. If the Contested Garment Principle were applicable for every allocation, the players could, in the event, attain the solutions above satisfying these properties.*

Proposition 4. *If the Contested Garment Principle cannot be valid for each allocation, as many experiments show, then the Mishnah solutions do not necessarily hold.*

9.4.5 The Loss Minimization Principle

"Two-armed bandit" is the jargon of the slot machine. The two-armed bandit problem is based on the choice between the two slot machines whose gain distributions are unknown to the participating players. This problem was argued by Bellman (1961) in the context of dynamic programming, by Rothschild (1974) in economics, and by Holland (1992) in genetic algorithm. Recently, economists have begun to reconsider this problem in the context of Holland's work. This problem shows that the traditional individualistic method of optimization does not work even in the Bayesian domain. In the two-armed bandit setting, agents – with their finite lifetime – cannot estimate the true distributions by experience or by observation. Applying these same settings to this example, it then holds that:

Proposition 5. *The Maximization Principle fails to work; only the Loss Minimization Principle survives as an available means to achieve performance.*[8]

Corollary 1. *A winner under uncertainty cannot be an individualistic optimizer. Individualistic optimization cannot necessarily be guaranteed.*

9.4.6 Adaptation in the Theory of Type Distribution

By virtue of the propositions above, it is hard to say that we can detect individualistically the best choice by learning under uncertainty where a microeconomic agent must always be exposed to loss. Because of this, a macroeconomic state will be comprised of at least two different groups, such as the successful group (for the best choice) and the failure group. Thus, in our economy, there may always be both winners and losers who are interacting, e.g., by imitation, or disguise. The "losers" must then imitate or learn from the "winners."

This is an evolutionary process of the economy, as Joseph Schumpeter suggested. An example is the competition of firms. It is not unusual to verify that the firms form sub-groups with a lower productivity, and less technical progress may survive even in the long run. In other words, the inferior agents could never be removed even at the limit.[9] In fact, we often see this kind of situation in the actual economy. Such an adaptive process in social interaction can be well defined by an appropriate stochastic model.

9.5 The Departure from Individualistic Rationality

In the earlier sections, the collapse of economic rationality as the general principle has been argued. Many economists believe even now that particular rationalities still work. Some refer to kinds of game theoretic rationality. In the preceding section, however, several particular rationalities and their failures were examined.

[8]See Holland (1992, Chap. 5). It no longer has the aesthetic beauty of dual principles of cost minimization and utility maximization.

[9]See Aoki and Yoshikawa (2006).

Rationality, either in general or in a special particular form, is not to be regarded as the so-called panacea in the complex socio-economic system.

9.5.1 The Victory of Patience

There may be various possibilities for agents under uncertainty, for example, the entry of a heterogeneous agent in the local market. In this market, an old chain store was operating for a long time. The store is the long-run player that behaves without regard to the short-run pay-off. Now suppose that an arriving chain store that is about to enter the market were a short-run player. The latter player cannot only attempt to execute his strategy all at once but can also repeat his move. Because of the short-run property, however, this arriving agent must find his final solution within a certain finite time. Thus we have the two kinds of heterogeneous agents, i.e., the short-run player and the long-run player.

The patience of the old chain store could successfully block the entry of the arriving agent, if the new agent believed the Bayesian guess. This game is called "the chain store game under uncertainty".

The weak chain store must accept and coexist with the arriving chain store. This strategy is called the "match strategy". There is, of course, a counter strategy, namely, the "price-cutting strategy". If the chain store had the next pay-off table, this store could employ the price-cutting strategy as the counter measure (Table 9.3).

The old chain store will be able to behave as if the table were realized for this store, because the arriving chain store does not really know whether this table is true, i.e., if the old chain store is strong or weak. Let the probability ε be the belief that it is strong, while the probability $1 - \varepsilon$ is the belief that it is weak. Here we set $\varepsilon = 0.05$, for instance. The manager can thus make his *ex ante* belief on the probability distribution. If the arriving chain store were faced with the same expected pay-off both for the price-cutting strategy and for the match strategy, then it should be indifferent between the entry and the retreat:

$$\pi \cdot -1 + (1 - \pi) \cdot 3 = \pi \cdot 0 + (1 - \pi) \cdot 0 \tag{9.6}$$

The probability satisfying this equation $\pi = 0.75$ is called the boundary probability. If the ex ante probability ε exceeds 0.75, namely,

$$\varepsilon \cdot -1 + (1 - \varepsilon) \cdot 3 < \varepsilon \cdot 0 + (1 - \varepsilon) \cdot 0, \tag{9.7}$$

Table 9.3 The pay-off matrix for the strong chain store

	The arriving store	
The old	Entry	Retreat
Match	(4, 3)	(8, 0)
Price-cutting	(6, –1)	(8, 0)

retreat must be chosen. As long as this holds and the *ex ante* probability is not revised, the arriving chain store cannot help giving up its entry.

Thus Fudenberg and Levine (1989) succeeded to prove the following proposition.

Proposition 6. *Let the short-run player make the ex ante belief* $\varepsilon \leq \pi$. *There exists the maximum number of times in the game that he can believe that the old store must be strong, if he continues to revise his belief according to the Bayesian rule.*

Hence the old but weak chain store could survive if it should continue to execute its price-cutting until the round of the game reaches the maximum M.[10]

Thus the patience of the long-run player, faced with the short-run and rational player, will lead to its being a winner, because the invader is a heterogeneous agent of the short-run player.

9.5.2 The Utilitarianism of Heterogeneous Interacting Agents

Usually, there will be a stochastic branching process of both agents where a typical individualistic optimization can stumble. The utilitarianism based on the building blocks of homogeneous producers and homogeneous consumers must break down. It deserves noting that utilitarianism by itself is not necessarily wrong. The utilitarianism of heterogeneous interacting agents is, in fact, necessary. This view however is not a new one; it can be traced back to the thought of the Mozi school in 500 B.C. in ancient China. Mozi had a definite idea of the heterogeneous agent, which is still applicable to the idea of multiple stakeholders in the present day.

An agent population of each type is called a sub-group. The types discussed here are a winner and a loser. The winner (or the sub-group of winners) that earns the highest gain at present may be imitated by the other remaining sub-groups failing to earn the highest gain.

The source of transition rates reflects the decision making of social interaction via sequential choices. Suppose there are a finite number of behavioral alternatives. Then there are imagined consequences, i.e., a sequence of anticipations of the possible decisions. In this circumstance, the multinomial logit type utility seems appropriate. The social interaction and coordination between the heterogeneous agents is a process of mutually interdependent communications. The utility of each agent is always affected not only by his own specific attribute and by others' behavior but also by the sub-groups' characteristics. In such an environment, the

[10] In our example, we set $\varepsilon = 0.05$, and calculated $\pi = 0.75$. It then the maximum number is 10.413 since

$$M \leq \frac{\ln \varepsilon}{\ln \pi} = 10.413.$$

neoclassical utility theory does not work, because the foundation of this theory originally is only limited to the purely individualistic attempts.

9.5.2.1 Weidlich–Helbing Formula

We then define the utility or preference of a decision on the alternative x for agent i whose last choice was the decision y:

$$U^i(x|y;t)$$

The probability for occurrence of the imagined consequence $\imath$, due to $f(\imath) = x$, will be

$$p^i(x|y;t).$$

According to Helbing (1995), the utility of the multinomial logit model is applied to the probability:

$$p^i(x|y;t) := \frac{e^{U^i(x|y;t)}}{\sum_{x'} e^{U^i(x'|y;t)}}. \tag{9.8}$$

Hence the probability that the next k arguments are all in favor of the alternative x will be as follows:

$$p'(x|y;t) = \hat{p}(x|y;t)^k = e^{k\hat{U}(x|y;t)}. \tag{9.9}$$

Here we employ the scaled preference:

$$\hat{U}(x|y;t) := U(x|y;t) - \ln \sum_{x'} e^{U(x'|y;t)}.$$

The *attractiveness* of the decision x for agent i who chose the behavior y will be expressed:

$$U^{i'}(x|y;t) := k\hat{U}^i(x|y;t) - \ln \sum_{x'} e^{k\hat{U}^i(x'|y;t)}. \tag{9.10}$$

By the use of $U^i(x,t)$, the utility of behavior x, and the concept of distance between behaviors x' and x, Helbing concludes *readiness*:

$$R(x'|x;t) := \frac{e^{U^i(x',t)-U^i(x,t)}}{D^i(x',x;t)}.$$

The elements for the decision on x by subgroup σ:
flexibility at time t:

$$\nu_\sigma(t).$$

distance between x and x' at time t:

$$D^\sigma(x',x;t) = D^\sigma(x,x';t) > 0.$$

effort at time t:

$$S^\sigma(x',x;t) := \ln D^\sigma(x',x;t).$$

Thus the social transition rate for a subgroup σ, w_σ can be summarized into

$$w^\sigma(x'|\,x;t) = \nu(t)R^\sigma(x'|\,x;t) \quad \text{for} \quad x' \neq x. \tag{9.11}$$

If our solution were inclined to be fixed in the long-run solutions, ν and R^σ could contribute to the trend of preference of the subgroup σ; attractiveness could make out the strength of preference for the subgroup σ.[11]

9.5.3 The Lottery System Can be Observed by the Urn Process

The complex economic system has the property to converge toward a biased distribution, which was often called a "disequilibrium cumulative process". Arthur (1994) characterized this process as increasing returns and path dependence. In this process, competition has been much changed.

The new ideas on the complex socio- and economic-system often were produced by applying the theory of the "urn" used in statistics and in population genetics. This suggests that the present socio- and economic-system has become a lottery system.

The Polya Urn Principle Arthur, when arguing the property of the reinforcement mechanism in general – not in spatial – interests, suggests the use of the "Polya Urn Process". The original urn process is the process of Polya and Eggenberger in 1923:

> Think of an urn of infinite capacity to which are added balls of two possible colors – red and white, say. Starting with one red and one white ball in the urn, add a ball each time, indefinitely, according to the rule: Choose a ball in the urn at random and replace it: if it is

[11]In order to argue these precisely, according to Weidlich (2000), we need to construct the socio-configuration by employing the personal variables and material variables pertinently specified, also defining the trend and control-parameters, and then the autonomous self-contained probabilistic sub-dynamics on the social system.

red, add a red; if it is white, add a white. Obviously this process has increments that are path-dependent – at any time the probability that the next ball added is red exactly equals the proportion. We might then ask: does the proportion of red (or white) balls wander indefinitely between zero and one, or does a strong law operate, so that the proportion settles down to a limit, causing a structure to emerge? [cited from Arthur (1994, p. 36).]

In a generalized version of the Polya Urn Process to allow for a greater number of agents (colors) than two, a random vector X whose components are proportions of each agent, is introduced instead of a numerical value (ratio).[12]

9.6 The Key Ideas for the New Economics

We can normally distinguish between the slow variables and the fast variables of a dynamic system. By referring to slow variables, the synergistic approach is used to find the order parameter to construct the master equation for studying the dynamic properties of a system. This approach never depends on so-called classical mechanics. In the new approach, each state could be described in a detailed balance of inflow and outflow, and the emergence of fluctuations of states can also be analyzed.

The stochastic evolution of the state vector can be described in terms of the master equation as equivalent to the Chapman–Kolmogorov differential equation system. The master equation leads to the aggregate dynamics, from which the Fokker–Planck equation could be derived. Thus we can explicitly argue the fluctuations in a dynamic system. These settings can indispensably be connected with the following key ideas, making feasible the type classification of agents in the system, and the tracking of the variations in cluster size. In such a framework, the equilibrium is a convergence of a statistical distribution and not a fixed point. The new method employs a set of statistical distributions of fractions of agents by types available in the new literature of combinatorial stochastic processes to reconstruct macroeconomics. The applications of this method are strenuously developed by Aoki (1996, 2002) and Aoki and Yoshikawa (2006).

Before the stochastic process for this argument is discussed, the two important ideas – of the exchangeability of agents and of the multiplicity of microeconomic configurations – must be understood.

Exchangeability between agent i and j As mentioned above, we cannot immediately determine which agent will achieve the best performance at a glance. This implies that an agent of a type i will be the agent of another type j in course of time. Here i can be exchanged with j. The idea of a master equation depends on this kind of exchangeability. It is noted that exchangeability may impose a double meaning on a state.

---

[12] See Theorem 3.1 of Arthur (1994, Chap. 10, pp.189–190).

Multiplicity of microeconomic configurations The total number of states is set N. The $n(t)$ indicates the state variable. There are $\frac{N!}{n!(N-n)!}$ ways to realize the same n/N. In other words, the equilibrium distribution gives the idea of multiplicity of micro-configurations that produce the same macro value.

9.6.1 The Exchangeable Agents in the Combinatorial Stochastic Process

The exchangeable agents come out by the use of a random partition vector in the idea of statistical physics or population genetics. The partition vector provides us with the state information.

Thus, the size-distribution of the components and their cluster dynamics can be argued.

> Define N as its most countable set, in which the probability density of transition from state i to state j is given respectively. In this setting, the dynamics of the heterogeneous interacting agents give the field where an agent can become another agent. It is also important to note that this way of thinking easily welcomes the unknown agents.

For example, in the financial markets, there are two different firms: the firm self-financed for investment, and the firm financed by banks. Focusing on firms of the same size capitalization for the two types, in this partition, the case of bankruptcy of the latter type firm as the result of bank loan diminution in the birth–death process of stochastic dynamics could be tracked.[13]

9.6.2 The Arrival of Unknown Agents

The population in human society does not consist only of age cohorts. Given the random partition vector and a new sampling formula, a new analysis of the social dynamics of multiple cohorts and their interactions can be developed. The distribution of order statistics on the size of groups can be calculated under a certain reasonable set of assumptions when the invariance of several groups of different types are known.

Ewens sampling formula (Ewens 1972) gives the invariance of the random partition vectors under the properties of exchangeability and size-biased permutation. The Ewens sampling formula is the case with one parameter. In order to deal with unknown agents, a new formula is required. *The Pitman sampling formula* can deal with the case of two parameters. Unknown agents often emerge, and a distribution of the emergence of unknown agents can be introduced according to the Pitman sampling formula (Pitman 1995 etc).

[13]The numerical simulation of this model continues to be developed by the research group of Mauro Gallegati. See for example Gallegati et al. (2006).

9.6.2.1 Pitman's Chinese Restaurant Process

There are an infinite number of round tables in the Chinese restaurant that are labeled by an integer from 1 to n. The first customer, numbered 1, takes a seat at the table numbered 1. Suppose that the customers from No.1 to No. k in turn take their seats at their tables from No.1 to No. k. Here the c_j customers take their seat at the j-th table. The new arrival comes out! The next arriving customer has two options: he or she can either take a seat at the $k+1$-th table by the probability

$$\frac{\theta + \kappa\alpha}{\theta + n},$$

or at the table j, one of the remaining tables ($j=1, \ldots, k$) by the probability

$$\frac{c_j - \alpha}{\theta + n}.$$

Here two parameters, θ and α, are used. The distribution of the state on how the customers take their seats is then given by the following formula:

$$\frac{n!\theta^{[\kappa:\alpha]}}{\theta[n]}\Pi_{j=1}^{n}\left(\frac{(1-\alpha)^{[j-1]}}{j!}\right)^{c_j}\frac{1}{c_j!}.$$

Here

$$\theta^{[j]} = \theta(\theta+1)\cdots(\theta+j-1); \theta^{[j:\alpha]} = \theta(\theta+\alpha)\cdots(\theta+(j-1)\alpha).$$

9.6.2.2 Ultrametrics

When a policy shift or an institutional change has taken place, there may be various effects on particular agents as well as effects on the whole population. In a simple case, the effects of mutual interaction can be measured by ultrametrics, that quantifies the distance between the different groups, thus obtaining the measure of interaction. Thus, by way of ultrametrics, there is a case where the effects of how a policy shift or an institutional change could enhance the welfare of particular agents and worsen the welfare of other agents can be assessed.[14]

[14]See Aoki (2002).

9.7 Concluding Remarks: A Deeper Consideration of the Utilitarianism of Heterogeneous Agents

As argued above, in particular in Sect. 5, it is clear that individualistic rationality must be abandoned. Instead, the utilitarianism of heterogeneous interacting agents is proposed. In this new framework, the solidarity formation among the heterogeneous interacting agents should be the most important matter.

9.7.1 Other-Regarding Preferences of the Human Mind

In economics or political economy, a selfish statement may not then be regarded as the revelation of the selfish gene. The personal selfish preference can be justified by a kind of mutualism in biology. The rational postulates on individualistic behavior will not necessarily guarantee the selfishness of the result of choice. Bowles and Gintis ask in the following manner:

> Is cooperation, then, just an expression of self-interest? Do other-regarding preferences such as a reciprocity and altruism play no role in the explanation of human cooperation? The answer hinges on whether most forms of cooperation are altruistic [in society] or mutualistic [in biology]? Bowles and Gintis (2005, p. 22)

Utility maximization by itself is not a presentation of selfishness, because the equipment to measure how much an individual's maximization affects others' utility is not installed in our individualistic utility function. In order to argue selfishness, our utility must be modified, as was considered in the earlier sections. What matters are the strength and the direction of preference. If we employ the multinomial logit type utility, we will use the information of many attributes of the individual like age, gender, and income.[15]

The strength of preference must be associated with the direction of strategy, because selfishness cannot be defined without the interactions between the agents. The direction of strategy will depend on the situation of the community as a whole.

9.7.2 Avatamsaka Game

In the Avatamsaka game, (Aruka 2001), selfishness would not be determined even if the agent selfishly adopted the strategy of defection (Table 9.4).

[15]"The analyst has to identify those that are likely to explain the choice of the individual. There is no automatic process to perform this identification. The knowledge of the actual application and the data availability plays an important role in this process" Bierlaire (1997, p. 4).

Table 9.4 Avatamsaka game

	Player B	
Player A	Defection	Cooperation
Defection	(0, 0)	(1, 0)
Cooperation	(0, 1)	(1, 1)

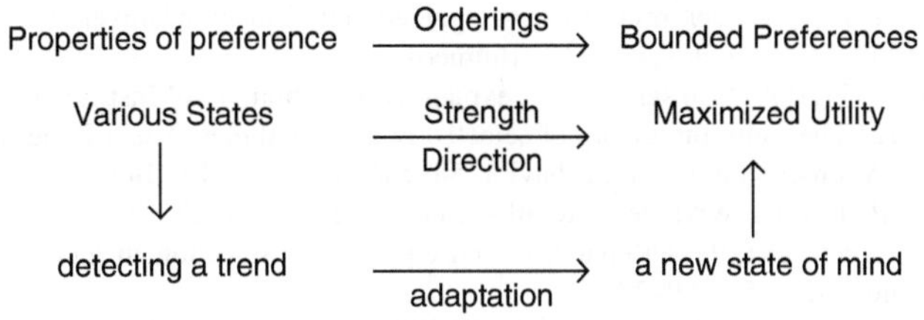

Fig. 9.4 Our new scheme on the utility function

This game belongs to a class of games in which the result for each agent will depend merely on the decisions of others, but not on his own decisions. The opposite is a one-person game (Robinson Crusoe's optimization problem), in which the decisions of others do not have any effect on a player's utility. In the Avatamsaka game, individual selfishness could only be realized if the other agent cooperates. Any certain gain from defection can never be assured by defection alone. The sanction by defection as a reaction of the rival agent never implies selfishness of the rival.[16]

The strength and direction of preference may thus be indispensable to formulate the evolution of utility. Preference may therefore be accompanied with the trend of selfishness. Selfishness may be a basic trend of the human mind (Fig. 9.4).

References

Akiyama E, Aruka Y (2006) Evolution of reciprocal cooperation in the Avatamsaka game. In: Namatame A, Kaizoji T, Aruka Y (eds) The complex networks of economic interactions: essays in agent-based economics and econophysics. Springer, Heidelberg, pp 307–320

Aoki M (1996) New approaches to macroeconomic modeling: evolutionary stochastic dynamics, multiple equilibria, and externalities as field effects. Cambridge University Press, Cambridge, New York

Aoki M (2002) Modeling aggregate behavior and fluctuations in economics: stochastic views of interacting agents. Cambridge University Press, New York

Aoki M, Yoshikawa H (2006) Reconstructing macroeconomics: a perspective from statistical physics and combinatorial stochastic processes. Cambridge University Press, Cambridge, New York

[16] As for the classification of dependent games, see Akiyama and Aruka (2006).

Arthur WB (1994) Increasing returns and path dependence in the economy. University of Michigan Press, Ann Arbor
Arthur WB (1996) Increasing returns and the new world of business. Harvard Bus Rev 74:100–109
Aruka Y (2001) Avatamsaka game structure and experiment on the Web. In: Aruka Y (ed) Evolutionary controversies in economics. Springer, Tokyo, pp 115–132
Aumann RJ, Maschler M (1985) Game theoretic analysis of a bankruptcy problem from the Talmud. J Econ Theory 36:195–213
Bellman R (1961) Adaptive control processes: a guided tour. Princeton University Press, Princeton
Bierlaire M (1997) Discrete choice models (mimeo)
Bowles S, Gintis H (2005) Can self-interest explain cooperation? Evol Inst Econ Rev 2(1):21–41
Ewens WJ (1972) The sampling theory of selectively neutral alleles. Theor Popul Biol 3:87–112
Frank R (2003) Microeconomics and behavior, 5th edn. McGraw-Hill, Boston
Frank R, Cook P (1995) The winner-take-all society. Free, New York
Fudenberg D, Levine D (1989) Reputation and equilibrium selection in games with a patient player. Econometrica 57:759–778
Gallegati M, Palestrini A, Delli Gatti E, Scalas E (2006) Aggregation of heterogeneous interacting agents: the variant representative agents framework. J Econ Interact Coord 1:5–20
Helbing D (1995) Quantitative sociodynamics: stochastic methods and models of social interaction processes. Kluwer Academic, Dordrecht
Holland JH (1992) Adaptation in natural and artificial systems. MIT Press, Cambridge, MA
Mainzer K (2005) Wass sind komplexe Systeme? In: Symposium zur Gr¨undung einer Deutsch-Japanischen Akadmie f¨ur integrative Wissenschaft, 2005 J.H. R¨oll Verlag, pp 37–77
Pitman J (1995) Exchangeable and partially exchangeable random partitions. Probab Theory Relat Field 12:145–158
Poser H (2005) The prediction problems in the complex sciences. In: Complexity and integrative science. Koyo-Shobo, Kyoto, pp 3–26 (in Japanese)
Rothschild M (1974) A two-armed bandit theory of market pricing. J Econ Theory 9:185–202
Schefold B (1989) Mr. Sraffa on joint production and other essays. Unwin Hyman, London
Schneider E (1934) Theorie der Produktion. Springer, Berlin
Weidlich W (2000) Sociodynamics: a systematic approach to mathematical modeling in the social sciences. Harwood Academic Publishers, Amsterdam (The Gordon and Breach Publishing Group). [Reprinted by Taylor and Francis (2002); Paper edition, Dover Publications (2006); Japanese translation, Morikita Shuppan (2007)]

Part III
Avatamsaka's Dilemma of the Two-Person Game with Only Positive Spillover

Chapter 10
Avatamsaka Game Structure and Experiment on the Web

Yuji Aruka

Abstract *Avatamsaka* is a well known *Mahayana Buddhist Sutras.* A Japanese professor, working in the field of Buddhists philosophy, skillfully illustrated the situation of heaven and hell in terms of *Avatamsaka* (Kamata 1988, pp. 167–168). Suppose that two people sit down at the table, across from each other. They are bound with rope so that one arm only is free, and are then each given a very long spoon. Own selves cannot be served by the use of a too long spoon. There are enough food for both of them on the table. If they cooperate and feed each other, they will both be happy. This is defined heaven. However, if the first is kind enough to provide the second with a meal, but the second does not feel cooperative, then only the second gains. This must give rise a feeling of hate in the first. This situation denotes hell. The gain structure does not only depend on an altruistic willingness to cooperate. On an individual level, there is no difference between cooperation and refusal and the same is true for risk taking. A solution for expected utility maximization gives infinite equilbria. Our interest is to find out a way to heaven from the other possible situations. This paper is concerned with how an actual player would react in such experiments.

Keywords Avatamsaka · Positive spillovers · Evolutionary game · Experiment on the Web

---

The project underlying this paper has been financially supported by the Japan Science Promotion Society. The first draft was written in Clare Hall, Cambridge UK in May. This paper was presented at J.-W. Goethe Universität for Frankfurter Volkswirtschaftliches Kolloquium (Frankfurt Economic Lectures) in 27 Nov 2000. I am especially grateful to Professor Bertram Schefold to provide me with his class and facilities for conducting experiment of 12 Dec 2000.

Y. Aruka
Faculty of Commerce, Chuo University, Higashinakano, Hachioji, Tokyo 192-0393, Japan
e-mail: aruka@tamacc.chuo-u.ac.jp

Y. Aruka (ed.), *Complexities of Production and Interacting Human Behaviour*,
DOI 10.1007/978-3-7908-2618-0_10, © Springer-Verlag Berlin Heidelberg 2011

10.1 Some Features of Avatamsaka Game

10.1.1 Road Example

We first set the *Avatamsaka Game* in a realistic example of a secular situation as the road example. We imagine "driving a car" in a busy place. When we wish to enter into a main street from a side road or a parking area along the road, it would be much better for us that someone can give way for us, although would not gain anything by giving way to someone they might never meet. This kindness, done by a stranger, is our gain. The symmetrically opposite situation in which we give way to a stranger the same relation. Anybody who does a kind deed for someone can never be guaranteed to gain from that other person. If nobody never gave way to anybody else, the situation on the roads would be attrocious. This support the old axiom,[1] *One good turn deserves another.*

It seems to me, however, we usually adopt a kind of mixed strategy rather than a pure strategy. Perhaps we rarely behave as extremists in any situation. This may be the result of our own *culture climate*, which develops as a consequence of the actions of agents living in the community.

10.1.2 Two-Person Game Forms

Here, the Heaven and Hell in Avatamsaka eare shown in a numerical example. The Avatamsaka game is a kind of *coordination game*.

This can be seen this situation in the Table 10.1, where

$$a = 0, b = 1, c = 0, d = 1.$$

This can easily be checked to see one of the Hell situations: if you chose "Cooperation" nevertheless the other "Defection", you lose your point. Put another way, you can get point by choosing Defection, if the other is altruistic.

Table 10.1 Coordination game and Avatamsaka game

(1) Coordination game		
Strategies	Player B	
Player A	Strategy 1	Strategy 2
Strategy 1	(a, a)	(b, c)
Strategy 2	(c, b)	(d, d)
(2) Avatamsaka game		
Strategies	Player B	
Player A	Defection	Cooperation
Defection	(0, 0)	(1, 0)
Cooperation	(0, 1)	(1, 1)

[1]This axiom is deeply meaningful, and should not be illustrated in terms of an individualistic utility function.

10.1.3 Infinite Equilibria under Expected Utility Maximization

We first state briefly how this game gives infinite equilibria in view of expected utility maximization. To do this we remove any equilibrium-selection mechanism in our framework. Under the previous conditions it was important to identify a multiple equilibrium with the others. The present problem unambiguously exits from this argument. The expected utility of Player A, when he uses Strategy D with probability q as well as Strategy C with probability $1 - q$, is:

$$E(D,q) = 0 \cdot q + 1 \cdot (1-q) = 1 - q,$$

if Player B always employs D. For the case Player B always uses Strategy C, his expected utility is:

$$E(C,q) = 0 \cdot q + 1 \cdot (1-q) = 1 - q.$$

It can be seen that any mixture of strategies can be called *a mixed strategy* if all the expectation happens to be equal. There indeed are infinite solutions q fulfilling

$$E(C,q) = E(D,q).$$

It is clear that the welfare results of this game are well ordered in a configuration of complete coordination, i.e., a combination of (Cooperation, Cooperation) as the top level, to a lower combination as the decisions to adopt Cooperation decrease, if a repeated game is introduced.[2]

10.1.4 No Complementarity to be Found Except for Positive Spillovers

In our game we cannot find any complementarity relation as is often seen in other coordination games. A *game is complementary,*[3] if a higher action in view of welfare by Player B increases the marginal return to the action of Player A. The situation can then be given a coordination feature. Define, first of all, a marginal return of action switch. If Player A can strictly switches his strategy from strategy 1

---

[2]This payoff matrix is

$$P = \begin{pmatrix} 0 & 1 \\ 0 & 1 \end{pmatrix}, P^t = \begin{pmatrix} 0 & 0 \\ 1 & 1 \end{pmatrix}.$$

It can be seen that the rank of the matrix is 1, i.e., either the rows or columns are linearly dependent. The solution of game then *degenerates*.

[3]This holds, for instance, if $b < a < d < c$. As for the definitions of Complementarity and Spillover see, for example, Cooper (1999).

to strategy 2, when Player B's action is strategy 1, the presence of an action switch can give the marginal return:

$$\Delta^{B:1}_{A:1\,\to\,2} = c - a > 0$$

If Player B adopted Strategy 2, the same switch may, for instance, give player A the higher marginal return[4]:

$$\Delta^{B:2}_{A:1\,\to\,2}(A) > \Delta^{B:1}_{A:1\,\to\,2}(A)$$

Welfare arrangements of choice combination are strictly ordered from higher to lower:

$$(d, d) \text{ is preferred to } (a, a).$$

In our Avatamsaka game, there may not be found any complementarity relation[5]:

$$\Delta^{B:2}_{A:1\,\to\,2}(A) = \Delta^{B:1}_{A:1\,\to\,2}(A)$$

On the other hand, the coordination game has usually another property called *positive spillovers*. This property implies that each player can increase his own reward as a result of the other's switch of action. In our setting, Player A can increase his reward by the other's action from Strategy 1 to 2:

$$\Delta^{B:2\,\to\,1}_{A:1}(A) = b - a > 0, \Delta^{B:2\,\to\,1}_{A:2}(A) = d - c > 0$$

Positive spillovers exist in Avatamsaka[6]:

$$\Delta^{B:1\,\to\,2}_{A:1}(A) = 1, \Delta^{B:1\,\to\,2}_{A:2}(A) = 1$$

Thus it can be seen that there are no complementarities, but there are positive spillovers in our Avatamsaka game.

10.1.5 How to Win the Game

In my teaching class experiences, I have noticed that many students hold a standard quite far from a supposed rational economic behavior: *a double standard* in a sense that there is a double binding action of the classical principle and the other one. We show a few examples. One is where students give the wrong answer when the

---

[4] $d - b > c - q > 0$ since $\Delta^{B:2}_{A:1\,\to\,2} = d - b$.
[5] $d - b = 1 - 1 = 0$; $c - a = 0 - 0 = 0$.
[6] $b - a = 1 - 0 = 1$; $d - c = 1 - 0 = 1$.

exercise asks them to decide how to impose a penalty to prevent a rival defecting in a *repeated* prisoners dilemma game. In this game,

$$c < a < d < b.$$

Students often favor the penalty of

$$(b - d) + (a - c) = b - c.$$

They should choose an answer which gives a positive result for themselves. They strongly dislike the difference $b - c$ between the players will increase, although they are only asked for a deterrent to defect. There are few students who impose the penalty of

$$(b - d)$$

as *a deterrent* to a rival's defection. Students or other subjects never stick to the classical principle of payoff maximization. They rather are adherents of the principle winning the game.[7]

It is clear that there are many cases does not conform to the expected utility maximization principle hypothesis. The above example shows that coordination cannot be guaranteed even though a complementary relation exists. *How to win the game is often irrelevant to the payoff maximization principle.* Players therefore maneuver some kind of spiteful action to attempt to maximize their *shadow points*. For example, *they choose an action which increase the difference in points between themselves and the other player as quickly as possible, and at almost every possible opportunity.* It can immediately be seen that this difference describes the negative spillovers in our terminology:

$$c - d.$$

In this context, our game may, in the first stage, result in the solution (Defection, Defection) because the solution can give the largest distance:

$$d - c = 1 - 0 = 1.$$

However this only produces all losers, and nobody gains anything, which contradicts the "winning" principle, and result in a "draw". Thus, the winning principle in this game needs to be more sophisticated. In our game, it seems very difficult for players to employ the winning principle, although they can use some spiteful action.

[7]See Chap. 5 of Taylor (1985), where such a discrimination principle is referred to. Another example from experimental results can be found in Cason et al. (1997).

10.1.6 Conversion of the Avatamsaka Game into a Coordination Game

Some economists who are familiar with the orthodox way of thinking may wonder whether this Avatamsaka structure deserves investigation by economists. They might suggest a transformation of the Avatamsaka structure into the usual coordination one. In order to achieve this, it is only necessary to introduce perturbation δ as is a very small number.

A perturbed version of case (1) may satisfy a complementary relation in a weaker form:

$$\Delta^{B:2}_{A:1\,\to\,2}(A) = d - b = 1 - 1 = 0$$

$$\Delta^{B:1}_{A:1\,\to\,2}(A) = c - a = 0 - \delta = -\delta$$

Hence

$$\Delta^{B:2}_{A:1\,\to\,2}(A) > \Delta^{B:1}_{A:1\,\to\,2}(A)$$

Another way to introduce perturbation may be considered, as in case (2) (Table 10.2). In this case, it also follows[8] that

$$\Delta^{B:2}_{A:1\,\to\,2}(A) > \Delta^{B:1}_{A:1\,\to\,2}(A)$$

These perturbed versions may create some trivial cases. This configuration may only give a dominant strategy either for Defect in case (1) or for Cooperation in case (2). No mixed strategy can be constructed in a meaningful way.[9]

Table 10.2 Perturbed Avatamsaka games

Case (1)		
Strategies	Player B	
Player A	Defection	Cooperation
Defection	$(0+\delta, 0+\delta)$	(1, 0)
Cooperation	(0, 1)	(1, 1)
Case (2)		
Strategies	Player B	
Player A	Defection	Cooperation
Defection	(0, 0)	$(1-\delta, 0)$
Cooperation	$(0, 1-\delta)$	(1, 1)

[8] $\Delta^{B:2}_{A:1\,\to\,2}(A) = 1 - (1-\delta) = \delta$ and $\Delta^{B:1}_{A:1\,\to\,2}(A) = c - a = 0 - 0 = 0$.

[9] In Case (1), for example, $E(D,q) = \delta\cdot q + 1\cdot(1-q) = 1-(1-\delta)q$ and $E(C,q) = 0\cdot q + 1\cdot(1-q) = 1-q$ together would require $\delta = 1$ if a mixed strategy did exist.

10.1.7 The Trembling Hands Example

We now turn to another perturbed version of the game which may be likened to *trembling hands* as caused by *emotional reasons*. On the one hand, a player may feel vexed when they offered cooperation and the other player defects. Again, we presume that this is a symmetrical situation. On the other hand, a player may feel pleased when they defect if the other player cooperates. We treat the feelings of vexation and pleasure symmetrically by the use of a small amount δ, which may be identified with pleasurable feeling, if δ is positive, and by $-\delta$ if the feeling is identified with vexation. We assume that they have a absolute value.[10] We now set $\delta > 0$, and thus we can consider the next proposition.

Proposition. *Suppose our payoff matrix of the perturbed version with human emotion to be:*

$$P = \begin{pmatrix} 0 - \delta & 1 + \delta \\ 0 & 1 \end{pmatrix}.$$

Players must employ the mixed strategy to use either Cooperation or Defection with probability 0.5 ½, whatever number is taken for δ.

Proof. Suppose Player A employs the strategies $p = (p_1, p_2)$, and Player B employs $q = (q_1, q_2)$. The value of this game on the payoff matrix P is then

$$v = q_2(p_2 + p_1(\delta + 1)) - p_1 q_1 \delta$$

Substituting $1 - p_2, 1 - q_2$ for p_1, q_1, we get

$$v = (p_2 - 1)\delta + q_2(1 - 2(p_2 - 1)\delta)$$

Player B's choice to maximize Player A's expectation is desired to be[11]

$$q_2 = 0.5.$$

This also holds for Player B. Henceforth, either player can use the strategy probability 0.5 to maximize his expected utility. □

This proposition may only suggest a technical possibility of a way out of our dilemma. It must not be overlooked that the *emotional reasons* might often be a means of settlement.[12]

---

[10]A utility function incorporating psychological relations is treated in Chap. 5 of Taylor (1985).

[11]Solve $\frac{dv}{dp_2} = 0$, given q. Since $\delta - 2q_2\,\delta = 0$, it follows that $q_2 = 0.5$. Similarly, $\frac{dv}{dq_2} = 0$ can be solved given p. There is a mixed strategy where both results are indifferent.

[12]In a different context, Frank (1988) argued for the virtues of human emotion. He said that emotions are especially useful or providing a fairly accurate measure of a person's intentions, because emotional responses are generally beyond a person's purposeful control, making them difficult to fake consistently and rendering a person's actions automatic and consistent.

10.2 Evolutionary Dynamics in Our Game

10.2.1 A Payoff Table Group Encounters

In the above proposition we have shown why a player can use this *mutant type* of mixed strategy which lies between a pure defect and a pure cooperation. Furthermore, it has the remarkable property unaffected by the magnitude of δ. This mutant when he encounter the representative mutant, as it lies between a pure defect and a pure cooperation, can at most earn its expected utility of 0.5. Now assign, for example, a probability

$$(\text{Defection, Cooperation}) = (0.8, 0.2)$$

It then holds that

$$0.5(0.8\cdot 0 + 0.2\cdot 0) + 0.5(0.8\cdot 1 + 0.2\cdot 1) = 0.5$$

which is equivalent to our representative mutant

$$0.5(0.5\cdot 0 + 0.5\cdot 0) + 0.5(0.5\cdot 1 + 0.5\cdot 1) = 0.5$$

There is a possibly infinite convex combination between pure defect and pure cooperation, but neither of them could gain more than our mutant strategy. I therefore believe that such behavior, i.e., to employ this even probability at each strategy, could be representative among all other mixed type strategies, as seen soon at the end of this paragraph. Even if this strategy predominated, the set of such strategies could not dominate the situation with the best solution (Cooperation, Cooperation) whose expected utility or welfare is always set 1.

In *an evolutionary game*, the higher the average welfare of the community average can be, the greater the growth rate of the group. It is interesting speculate whether a generation of mutant mixed strategy would evolve into an increasingly dominant influence in the community. The payoff structure of our evolutionary game, or a table of the payoff outcome of group encounter, is summarized in Table 10.3.[13]

[13]Defect *vs*. Defect implies that the resultant points for the D-group become 0 with probability 1, namely, $1\cdot 0 \rightarrow 0$. Defect *vs*. Cooperation implies that the resultant points for the D-group become 1 with probability 1, namely, $1\cdot 1 \rightarrow 1$. Defect *vs*. Mutant implies that the resultant points for the D-group become 0.5 since the player is defected with probability 0.5 and is cooperated with probability 0.5, namely, $0.5\cdot 0 + 0.5\cdot 1 \rightarrow 0.5$. Similarly it follows for the C-group: Cooperation *vs*. Defect implies $1\cdot 0 \rightarrow 0$. Cooperation *vs*. Cooperation implies $1\cdot 1 \rightarrow 1$. Cooperation *vs*. Mutant implies $0.5\cdot 0 + 0.5\cdot 1 \rightarrow 0.5$. It also follows for the Mutant-group: Mutant *vs*. Defect implies $0.5\cdot 0 + 0.5\cdot 0 \rightarrow 0$. Mutant *vs*. Cooperation implies $0.5\cdot 1 + 0.5\cdot 1 \rightarrow 1$. Mutant *vs*. Mutant implies $0.5\cdot 0 + 0.5\cdot 1 \rightarrow 0.5$.

Table 10.3 Payoff structure of evolutionary game

Species	*D*-group	*C*-group	Mutant
D-group	0	1	0.5
C-group	0	1	0.5
Mutant	0	1	0.5

This table shows the remarkable property that all the groups have the same payoff. It is natural that every group should have expected to survive and also that they should believe in their own belief's rationale. The initial states may all be unchanged and in this case, any *shuffle* action would not be useful.

Finally, it must be noted that this idea may be still kept, even if some variety of mixed strategy is to be introduced. Suppose there are three ways of mixed strategy. We can, for example, classify three groups according to the magnitude of the probability that they will employ D. If a player employs with a probability of 0.8 on *average*, they belongs to the D-biased-group. If this probability is 0.5, they belongs to the neutral group. If its probability is 0.2, they belong to the C-biased-group. This player then has the same payoff relationship with any group.[14] This situation does not generate any motion because these interaction take palce without any perturbation. We now consider the even probability case of a mutant as *a representative* mixed strategy groups.

10.2.2 Modified Tables for Evolutionary Dynamics with a Reinforcement Mechanism

Table 10.3 only gives a relationship of complete dependency between groups, since all depend on the initial values of their positions, and this does not change once the initial values were implemented. However, there is the possibility of motion in the *psychological reaction* of a player to the defection or cooperation of his opponent. This hypothesis can give some remarkably interesting dynamics. Taking this into account, we consider the situation given in Table 10.4.

Note that δ is only *a psychological factor*. The total points gained never contain an additional δ. However, δ implies a *reinforcement* for the player to continue to employ a strategy. A premium δ may encourage a player either to continue their last action or to attract the current action. We can design that each player is able to observe the list of all the current points of the other players. It is assumed that players can only decide their strategy by taking account of all the total points of others, i.e., the community average without any premium. There may also be a case which is so sophisticated that every player decides their strategy by calculating the community average including a premium.

[14] If a different player employs a D-biased strategy, he expects to earn $0.2(0.2 \cdot 1 + 0.8 \cdot 0) + 0.8(0.2 \cdot 1 + 0.8 \cdot 0) = 0.2$ when he is faced to a member of the same D-biased group, and so on.

Table 10.4 Payoff relationship of revised evolutionary game

Species	*D*-group	*C*-group	Mutant
D-group	$-\delta$	$1+\delta$	0.5
C-group	0	$1+\delta$	0.5
Mutant	0	1	$0.5+\delta$

We now formulate *a population dynamics with a reinforcement mechanism.* Let the payoff matrix of the community interaction to be A, its row to be a_i, and each proportion of the community population employing each strategy be x_i, where x_1 represents the ratio of the defect group to the community population, x_2 is the ratio of the cooperation group, and x_3 is the ratio of the mutant group. Denote

$$x = \begin{pmatrix} x_1 \\ x_2 \\ x_3 \end{pmatrix}$$

and let its transpose x^t. Here, $\sum_{i=1}^{3} x_i = 1$.

There are three ways to describe the community interaction matrix. A player who employs D may feel either vexed when defeated or pleased when winning.[15] The D-group then is faced with the following matrix:

$$A^D = \begin{pmatrix} -\delta & 1+\delta & 0.5 \\ 0 & 1 & 0.5 \\ 0 & 1 & 0.5 \end{pmatrix}, \; a_1^D = (-\delta, 1+\delta, 0.5).$$

A player employing C may be reinforced by his own feelings about premium δ when they adopt C against the same group. The C-group matrix is

$$A^C = \begin{pmatrix} 0 & 1 & 0.5 \\ 0 & 1+\delta & 0.5 \\ 0 & 1 & 0.5 \end{pmatrix}, \; a_2^C = (0, 1+\delta, 0.5).$$

Similarly, a player employing M may be reinforced by their own feeling of premium δ when adopting M against the same group. The M-group matrix is

$$A^M = \begin{pmatrix} 0 & 0 & 0.5 \\ 0 & 1 & 0.5 \\ 0 & 1 & 0.5+\delta \end{pmatrix}, \; a_3^M = (0, 1, 0.5+\delta)$$

The welfare of a group whose players decide the group's strategy can then be expressed as $a'_i A^i x$, and the average welfare as $x'_i Ax$, where $a'_i x$ is the i-th group's

[15]When a player is faced with a mutant player whose strategy is the even probability policy, premium may be offset, on average, as zero.

scores or welfare, and $\frac{\Delta x_i}{x_i}$ is the i-th group's growth rate on population. Our simple dynamics show that a relative population of the group increases if the group welfare is greater than the community average welfare. Thus our evolutionary dynamics equations can be stated as[16]

$$\Delta x_1 = x_1(a_1^D x - x^t A^D x),$$

$$\Delta x_2 = x_2(a_2^C x - x^t A^C x),$$

$$\Delta x_3 = x_3(a_3^M x - x^t A^M x).$$

10.2.3 A Motive for Metamorphosis

Group population dynamics works through an adjustment mechanism such that the group with the highest growth rate whose score is higher than the average community score can attract players from the other remaining groups, or in other words, induce them to change from their original strategy to the strategy of the group with higher scores. The group population dynamics alone does not represent any individualistic willingness, which is based on expected utility maximization.

We assume that each player has an equal probability and the probability that the i-th player will meet one particular group depends on the current scores of the groups. The motive for a change in action mainly is related to how large scores are distributed among groups. A group formation can be caused by such a kind of metamorphosis from C to D, and C to M; D to C, and D to M; M to C, and M to D.

Two forces for a dynamic motion can therefore be seen in our revised population dynamics. The C-group is attracted into the D-group if the D-group's *current* score is greater than that of the C-group. *The marginal preferences* of the groups may be ordered hypothetically according to the current distribution of the proportions of the groups in the community, i.e.,

$$(x_1, x_2, x_3).$$

On the other hand, the i-th group must reinforce into itself if its profit is above average:

$$a^i{}_i x - x^t A x.$$

Thus, every product of

$$x_i(a^i{}_i x - x^t A x)$$

[16]Alternatively, we may use a more sophisticated system by replacing A with A^i. A computer run of this sophisticated system $\{a_i^i x - x^t A^i x\}$ may produce essentially similar results because it shares with the same cumulating points.

may make an instant net difference between the exit or entry of each group, i.e., *metamorphosis*. We call a psychological factor *asymmetrical* if D has a payoff of $-\delta$ against D. Furthermore, if D has also a payoff of $1 + \delta$ against C, we call *symmetrical*.

When psychological effects are introduced, *a complementary relationship*, as defined in the previous section, whether asymmetrical or a symmetrical.

10.2.4 Population Dynamics in a Triangle

In our dynamics, we can always discover whether a mutant group will eventually either disappear or dominate. Extreme points in the unit simplex of probability support show the dominations of a single group, with the two extreme points being two groups trying to remove the other group. This dominant group stands on a base defined by the D-group extreme point and the C-group point.

Theorem[17]**.** *The strategies $x = (x_1, x_2, x_3)$, representing the ratios of D, C, and M, are all in a unit simplex. A player of M only employs their mutant strategy randomly with even probability. Suppose that there is a psychological complementary relation characterized by δ in the above definition.*

1. When metamorphosis from M to C works well, and if the initial starting point is appropriately chosen, there may be guaranteed path towards an increase C ratio and a fixed point with M exhausted.
2. There may be a large enough increase in the magnitude of the psychological factor δ to reverse metamorphosis from C to M.
3. The simultaneous increase of x_1 and x_2 , coevolution of C and D come from the metamorphoses of M.
4. The extremes D and M cannot be attracting points.

Let the *Liapunov Function* be

$$V(x) = x_1^{p_1} x_2^{p_2} x_3^{p_3}.$$

It then follows:

$$\dot{V}(x) = V(x)(p_1 \frac{\dot{x}_1}{x_1} + p_2 \frac{\dot{x}_2}{x_2} + p_3 \frac{\dot{x}_3}{x_3}).$$

If $\dot{V}(x)>0$ is always confirmed, any vector x shall asymptotically converge to a fixed point $x^* = (p_1, p_2, p_3)$ (Arrow-Smith and Place 1982, pp. 207–210).

[17]The proof of this theorem can be seen in Aruka (2000).

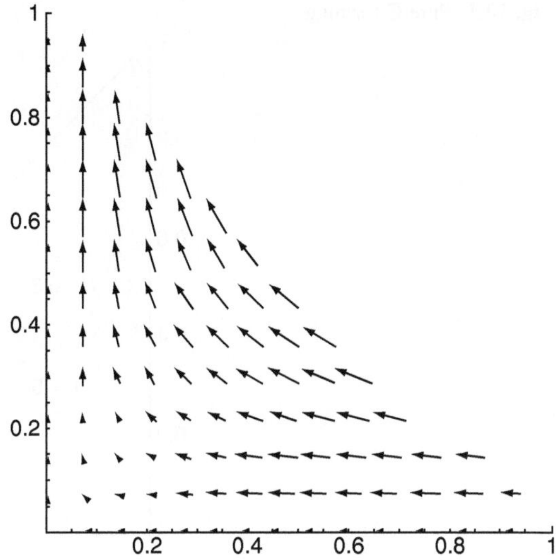

Fig. 10.1 Asymmetrical case of psychology

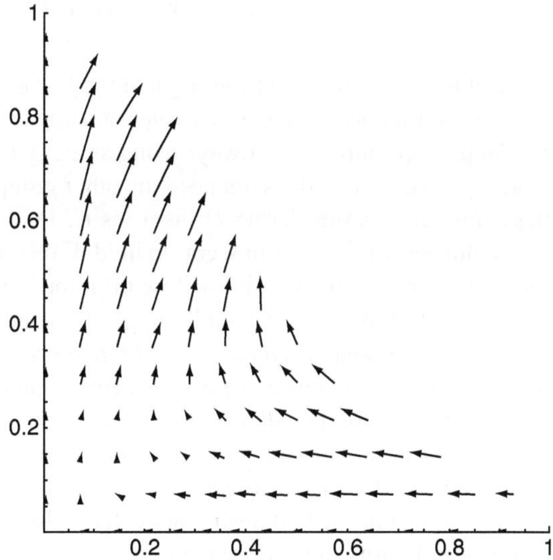

Fig. 10.2 Symmetrical case of psychology

We now straightforwardly illustrate the above statements by producing Figs. 10.1–10.3. Figure 10.1 shows a phase diagram of dynamic motion based on psychologically asymmetrical matrices, while Fig. 10.2 is a phase diagram based on the psychologically symmetrical matrices. Figure 10.3 shows in the system $\{a_i^i x - x^t A^i x\}$ a convergent trajectory towards pure C, provided that the order of magnitude of δ has been changed from 0.01 to 0.1.

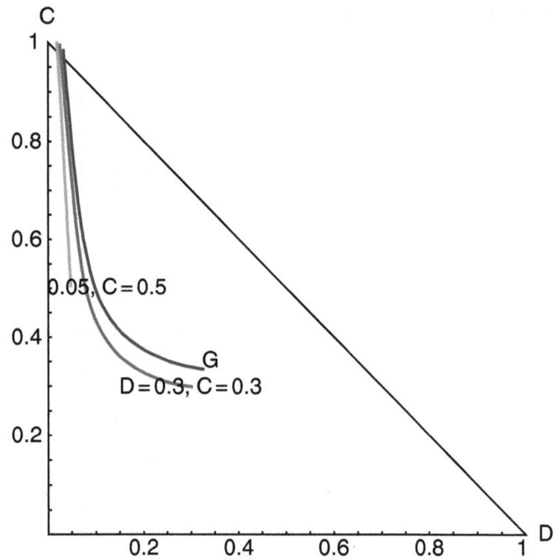

Fig. 10.3 Pure C forming

10.2.5 Effect Generated by an Intelligent Group

We end this section by considering that the *Volterra–Lotka Relation does not hold.* For the D-group to survive as a *winner* presupposes at least the positive ratio of the C-group. Then pure C, or always using strategy C will eventually die out. For the C-group to survive, it does not need the other group since it is achieving the best for the community. Considering D vis a vis C, D is *a predator* while C is the *prey*: a predator–prey relationship seems to hold. If this relation lasted forever, the growth rate of two species would give a close orbit the Volterra–Lotka equation. However, there are differences between this supposition and ours. *The difference here is that a predator may temporarily alter itself into a prey; in other words, D becomes C for a while.* This is why our game will not give regular fluctuations in the group ratios, and much more complicated behavior may be anticipated.

We have used a mutant strategy in the sense that player employs either D or C simply *randomly*. Any discretion by using an *intelligent* intention was never presumed. Now suppose that a very tactically skillful player to penetrates an opponent strategy and gains over the average of the community. Such a player must occasionally use a policy for "Tit for Tat",[18] but their personal ability never propagates among others, unless *a coalition group* organized by the smart player is permissible. Such an intelligent mutant will not realistically survive. So far in our game, skillful conjecture did not matter for the M-group. Suppose that a skillful man has a payoff matrix as in Table 10.5. Notice that a psychological score "over 0.5" must

[18]Even in this case, the player cannot keep 1 point, because their behavior might induce someone else to resist and exert D against them.

Table 10.5 Payoff for a skillful man

Species	*D*-group	*C*-group	Mutant
Mutant	0	1	over 0.5

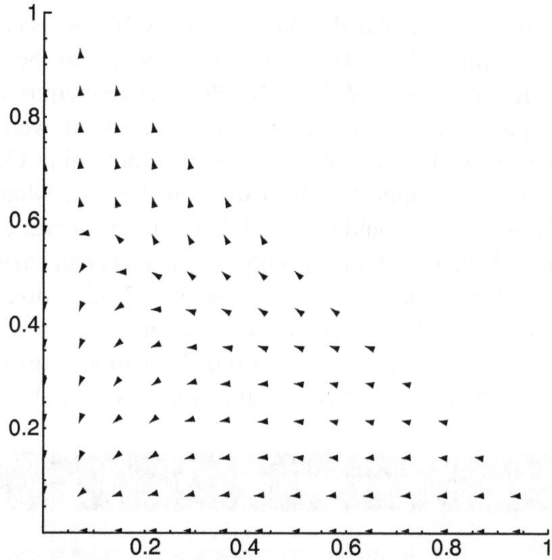

Fig. 10.4 Intelligent case

be realizable. The opponents who are facing the intelligent group must face a loss in their payoff. In details, it follows that

$$A^D = \begin{pmatrix} -\delta & 1 & 0.5-\delta \\ 0 & 1 & 0.5-\delta \\ 0 & 1 & 0.5+\delta \end{pmatrix}, \quad A^C = \begin{pmatrix} 0 & 1 & 0.5-\delta \\ 0 & 1+\delta & 0.5-\delta \\ 0 & 1 & 0.5+\delta \end{pmatrix},$$

$$A^M = \begin{pmatrix} 0 & 1 & 0.5-\delta \\ 0 & 1 & 0.5-\delta \\ 0 & 1 & 0.5+\delta \end{pmatrix}.$$

If the intelligent player should form their own coalition or group to attract other players, it would be easy to show that this group's rapid increase would soon reach a critical proportion. See Fig. 10.4. There would be different cases in the final state depending on initial conditions, e.g., (1) possibly all the mutants in the population, or (2) all the *C*-group, and so on.

Our *hypothetical* population dynamics could accumulate *a limiting sequence* towards a special base or an extreme point in whatever direction it turned. Some types of *degeneration* have been observed in our computer runs. It is interesting to consider how, and under what conditions, an experiment done with the human subjects would show a similar result. *The aim of this research, however, was to test statistically the data on the ratio of groups obtained by experiments with the various different samples.*

10.3 Experiment on the Web

10.3.1 Experiment on the Web

This experiment had a maximum ten subjects, as is standard practice in experimental economics (Fig. 10.5). The experiment may be found on the web by the use of software program *File Maker Pro.*[19] A near equivalent of the display layouts for this experiment in my server computer, which were developed by myself, will be reproduced on the Web by a well functioning CGI loaded in File Maker. It may not be not important to a user on the web what operating system is employed. However, it should be noted that only my server computer operates with the use of File Maker, and a maximum ten subjects can easily and simultaneously access my IP address via an internet browser.[20] Notice that File Maker Pro. A maximum of ten personal computers are sufficient if each has an internet browser installed, such as Internet Explorer 4.0 or later to support CSS1 (Cascading Style Sheet 1). Unfortunately, Netscape Navigator is not suitable at present.

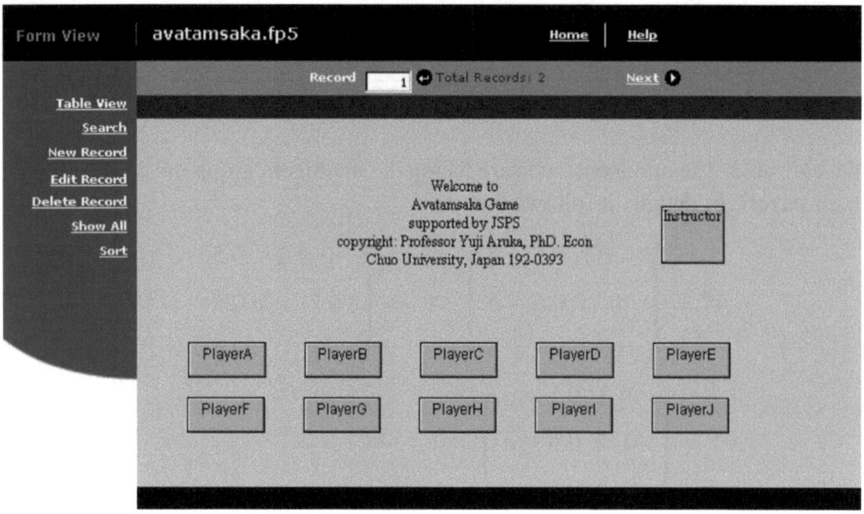

Fig. 10.5 Cover page of experimental program

---

[19]The advantage of data base software File Maker Pro *ver.5* is that "Web Companion", a very useful function of CGI (common gateway interface), is loaded.

[20]There are possible combinations with an even number of 2, 4, 6, 8, and 10, since our experiment is designed for a two-person game. Two persons are required for each match. If there are ten subjects, one must then play a match with the remaining 9 persons. We must therefore in total arrange the matches of $(10/2) \times (10 - 1) = 45$. This game is then finished. Incidentally, our program is also valid for all the standard two-person games such as the prisoners dilemma simply by revising the coefficient of payoff matrix.

10.3.2 Evolution of Strategies

In a computer simulation or in the experiment with human subjects, our aim was to investigate how a three groups, such as a pure *Defection*, a pure *Cooperation*, and a *Mutant* group $\{D, C, M\}$ as dealt with in the previous sections, would fluctuate. Different sequences were used in each sample to sample in the experiment in order to examine what degree of variance was due to different samples. Such samples may be domestic or international.

10.3.3 Monitoring the Others

Following convention, one exercise was given the participants at first. We confined the series of game of a maximum of ten subjects for ten successive times. Every subject could have as much information as his rivals. They can always monitor a current point of the other players, as well as their names. This partly takes into account the assumption of learning, which is discussed below.

10.3.4 Two Kinds of Experiment

In this experiment, it is also interested in checking whether *a pecuniary motive* would affect a player's action or not. Two different games, with and without payment had to be designed. Experiments in economics are usually used to test the pecuniary motives of economic behavior. Thus an experimental economist can generally reward each player's effort. In our game, we wish to control the experiment rather differently. We therefore arrange two types of experiment. One experiment can follows the traditional method of experimental economics of paying each player according to their achievement. The other is without payment. Our intention is to test statistically whether the motive during our game was biased to a pecuniary advantage or not.

10.3.5 Repeated Games

Our Experiment needs ten subjects and an instructor. Each subject is requested to play a two-person game within the rules, which appears in the computer program. One game may consist of a series of matches to be played with the other players. When there are ten subjects, each game consists of nine matches for each player. The players can run their game by referring to the points distribution of the whole community. This points distribution may also be reproduced in each game display. The player will repeat the same game ten times. Consequently, each player will play

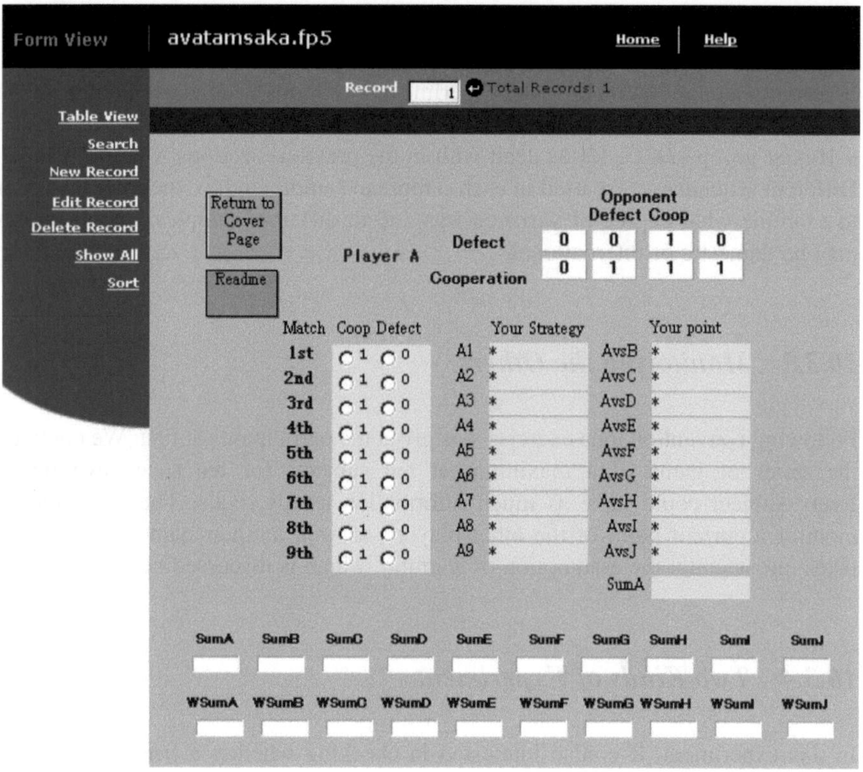

Fig. 10.6 Display for Player A

90 matches. The maximum possible number of points a player can receive is 90 points. On the other hand, the worst possible may be 0 (Fig. 10.6).[21]

10.3.6 Our Initial Concerns

In our original Avatamsaka Game scheme without the introduction of perturbation, only *positive spillovers* were implemented. The individual's motives, whether egoistic or altruistic did not work well owing to the entire lack of *complementary relation* in a payoff structure. We are concerned whether or not *positive spillovers* isolated from complementarities could still induce behaviors based on utility maximization.

[21]Instructor: A computer expert will be necessary to assist students to operate Internet Explorer or a similar browser. Subject must give comment: A final request will be given in the following form: Please give us a comment on the Avatamsaka game after finishing this game. Please use this line to enter the mail-sending form or equivalently [questionnaire] on the left frame. This comment will automatically be sent by the mail form system written in CGI.

In an iterated game, *learning effects* are one of the most important and controversial topics. It is usually assumed that without any repetition, the subjects cannot know the distribution of types on players in advance. By iterations of the same game, the subjects should learn this distribution. Even if there could be assured a *mathematically* symmetrical pay-off matrix, by which a unique solution is guaranteed, nevertheless, it would take too a long time, possibly an infinitely long time, to know the true effects of the learning thus experimentally revealed. Gale (1960) has pointed out this over 40 years ago, when he gave a classical proposition on learning in a two-person game.[22] Fortunately, our simulation results show that some sequences reach a stationary situation in a finite number of steps. We therefore restricted the number of trials of our game to ten, and these experiments will of course be shorter compared with computer simulation. But if some tendency towards an eventual situation is to be discernible, the data observed can deserve statistical testing.

Furthermore, there have been many investigations into whether or not an agent really learn to behave in a way which will increase their payoff. Testing learning effects is still an active area of experimental economics. However, we do not especially want to focus on any stochastic process of convergence by learning, and/or to ask whether human being is bounded rational or not. These are not our concerns here. We are interested in a question: *How can an emotion to cooperate be generated in concert?* In fact, however, we have introduced a kind of learning process. In order to achieve a well-organized experiment on this problem, we shall uncovered a range of types, by informing the subject of the gains of others. If a whole composition of the ratios of the groups always were monitored by each player, and at least partly detected by them knowing the current scores of the other players, they could each react unambiguously according to their own deductions, which are closely related to their emotional feelings. In our experiment, the emphasis is placed on taking into consideration the community trend, but not on individualistic learning by collecting information.

10.3.7 Experiment in Frankfurt am Main of December 2000

Finally, we mention the latest result of *Avatamsaka* experiment held in Germany.[23] These are given in Table 10.6, with no statistical test or any evaluation of their consequences.

[22]A convergence element might be too slow to deserve practical consideration, even if a convergence were guaranteed. See Gale(1960, pp. 250–256).

[23]This experiment on the web was held at the PC- Pool, J.-W. Goethe Universität (Frankfurt University), Frankfurt am Main from 11:00 to 12:00 hours, on 12 Dec 2000. The subjects were supplied from the class of Professor Bertram Schefold, Faculty of Economics, and Herr Roland Goslar, his assistant, who conducted the experiment by accessing the author's computer server in Tokyo, Japan. This experiment belongs to a class without explicit payment, although non-pecuniary prizes of cookie were allocated among subjects according to their resultant scores.

Table 10.6 Avatamsaka experiment in December 2000, in Germany (points and C-ratio)

(1) Points distribution										
	A	B	C	D	E	F	G	H	I	J
Round 1	9	6	7	6	8	7	7	7	6	7
Round 2	6	7	4	5	7	6	7	7	7	6
Round 3	7	7	5	5	7	7	6	5	5	6
Round 4	7	6	7	8	7	8	7	7	9	6
Round 5	9	8	8	7	8	8	7	7	8	7
Round 6	9	9	7	8	8	6	6	8	8	7
Round 7	8	3	6	6	7	7	4	8	8	6
Round 8	8	4	6	5	8	7	6	8	8	5
Round 9	8	8	7	7	8	6	6	3	7	6
Round 10	5	5	5	4	6	4	4	5	4	5
Total	76	63	62	61	74	66	60	65	70	61
Average	0.84	0.7	0.69	0.68	0.82	0.73	0.67	0.72	0.78	0.68
(2) Times that strategy C was employed										
	A	B	C	D	E	F	G	H	I	J
Round 1	8	8	9	3	7	7	8	9	4	6
Round 2	8	8	9	2	8	5	6	0	4	9
Round 3	6^{NA2}	8	9	6	7	7	6	0	5	6
Round 4	9	8	9	7	9	7	8	3	7	5^{NA1}
Round 5	9	8	9	8	8	5^{NA1}	5	9	9	7
Round 6	9	8	9	6	8	8	3	9	8	7
Round 7	9	8	9	6	8	8	7	7	9	6
Round 8	8	8	7	6	8	8	6	0	8	6
Round 9	8	8	6	0	9	8	5	9	8	4
Round 10	9	8	7	0	0	4	4	0	9	5
Total	83	80	83	44	72	67	58	46	71	61
Average	0.92	0.89	0.92	0.49	0.80	0.74	0.64	0.51	0.79	0.68

The superscript ^{NAx} shows that the number of *NA*(not available) values is *x*

References

Arrow-Smith DK, Place CM (1982) Ordinary differential equations. Chapman and Hall, London, pp 207–210

Aruka Y (2000) Avatamsaka game structure and it's design of experiment. Working Paper No.3, The Institute of Business Research, Chuo University, pp 15–18

Cason T, Saigon T, Yamato T (1997) Voluntary participation and spite in public good provision experiments: international comparison (mimeo)

Cooper RW (1999) Coordination games. Cambridge University Press, Cambridge

Frank R (1988) Passions within reason: the strategic role of emotions. W. W. Norton, New York

Gale D (1960) The theory of linear economic models (reprinted). University of Chicago Press, Chicago, pp 250–256

Taylor M (1985) Possibility of cooperation. Cambridge University Press, Cambridge

Chapter 11
Avatamsaka Game Experiment as a Nonlinear Polya Urn Process

Yuji Aruka

Abstract We imagine "driving a car" in a busy place. When we wish to enter into a main street from a side road or a parking area along the road, it must be much better for us that someone can give way for us although someone cannot gain any more by giving us whom someone might never meet. This kindness done by someone implies our gain. A symmetrical situation in replacing someone with us holds the same relation. Anybody who gives a kind arrangement to someone, can never be guaranteed to gain from another. If nobody never gave way for someone, the whole welfare on the roads should be extremely worsen. We call such a situation *Avatamsaka Situation* in the sense of the footnote.[1] The gain structure will not essentially depend on the altruistic willingness to cooperate. The model after *synchronizing* our original setting *albeit* a vulgar conversion can be written down as a variant of *coordination games*, although the model can then provide with infinite equilibria in view of individualistic expected utility maximization. A profile of strategic deployments on a set of $\{cooperation, defection\}$ may be argued in a repeated form. Thus one way to look for an eventual end point in our repeated game is to formulate the model in terms of *evolutionary game* of a simultaneous differential equations system (Aruka 2000, 2001). On the other hand, this paper will deal with a *stochastic* aspect of this evolutionary process. Finally, it is noted

[1]*The Original Source*: *Avatamsaka* is a well-known one of the Mahayana Buddhist Sutras. Shigeto Kamata, Professor Emeritus, the University of Tokyo, who is working under the field of Buddhists philosophy, he skillfully illustrated the situation of Heaven and Hell in view of *Avatamsaka*. See Kamata (1988, pp. 167–168). Suppose that two men *vis a vis* sat down at the table, across from each other. They however are tied with rope except for one arm only, then given each a too long spoon. Own selves cannot serve them by the use of a too long spoon. There are meals enough for them on the table. If they cooperate in concert to provide each other with meal, they can all be happy. This defines the *Heaven* or *Paradise*. Otherwise, one will be so kind to provide the other with meal but the other might not have a feeling of cooperation. This case however pays the other. This must give rise a feeling of hate for the other. This describes a situation enough to note the *Hell*.

This project was supported by *Japan Science Promotion Society*:Grant-in-aid for Scientific Research on Priority Areas (B) No.10430004.

Y. Aruka
Faculty of Commerce, Chuo University, Higashinakano, Hachioji, Tokyo 192–0393, Japan
e-mail: aruka@tamacc.chuo-u.ac.jp

Y. Aruka (ed.), *Complexities of Production and Interacting Human Behaviour*, 223
DOI 10.1007/978-3-7908-2618-0_11, © Springer-Verlag Berlin Heidelberg 2011

that we have *internationally* achieved *Avatamsaka Game Experiment on the Web* by my original program by he use of *File Maker Pro*'s web companion.[2]

Keywords Avatamsaka game experiment · Nonlinear evolution · Path dependent · Polya urn process · Positive spillovers

11.1 Characteristics of Avatamsaka Game

11.1.1 Synchronization

We can depict a picture for a vulgar image of *Avatamsaka* situation in terms of *synchronized* payoff structure for playing agents in terms of the two persons game form. At the beginning of our experiment, as we in fact used, we may suggest our subjects to remind of the next scenario:

- Only a piece of block cannot support a dish. If we have two pieces, they can manage to support a dish. Now you join a game.
- Your strategy is restricted either to put a piece beneath on the dish i.e., *cooperation* or not to put i.e., *defection.*
- Both players must move simultaneously.
- You must play according to *the next rule*: Both players cannot get a point by the move of "do not put". You can get a point by the move of "do not put" in case your move consequently never break the dish, but you cannot get a point by the move of "put" when the opponent did not put. You can get a point by the move of "put" when the opponent does put… (Fig. 11.1).

11.1.2 A Two Person Game Form

Now we can give the Heaven and Hell in *Avatamsaka* a numerical example. the Avatamsaka game is a kind of *coordination game*. You can see this situation from the Table 11.1, if it is set

$$a = 0,\ b = 1,\ c = 0,\ d = 1.$$

Specifically, see Table 11.2.

You can easily check to see one of the Hell situations: if you chose "Cooperation" nevertheless the other "Defection", you should lose your point. Put another way,

[2]We have already had experiments for several times at Chuo University, Japan, Frankfurt University, Germany, and Maquette University, USA.

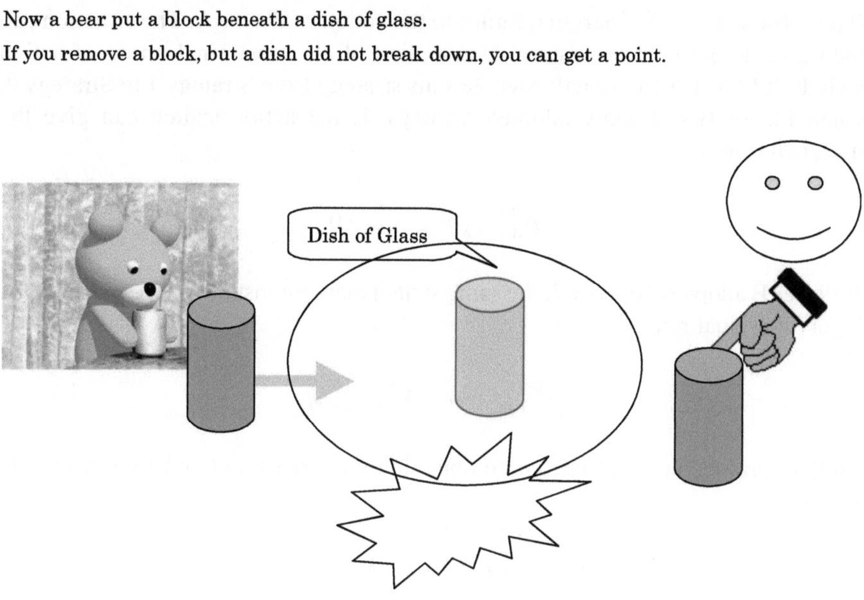

Fig. 11.1 An image of Avatamsake game

Table 11.1 Coordination game

Strategies	Player B	
Player A	Strategy 1	Strategy 2
Strategy 1	(a, a)	(b, c)
Strategy 2	(c, b)	(d, d)

Table 11.2 Avatamsaka game

Strategies	Player B	
Player A	Defection	Cooperation
Defection	(0, 0)	(1, 0)
Cooperation	(0, 1)	(1, 1)

you can get gain by choosing Defection, if the other has a kind of altruism. It may be easily verified that this game has infinite equilibria.[3]

No Complementarities Except for Positive Spillovers to be Found: In our game we cannot find any complementarities relation as often seen in many coordination games. It is called *a game is complementary*,[4] if a higher action in view of welfare by

[3]The expected utility of Player A, when he uses Strategy D with probability q as well as Strategy C with probability $1-q$, is $E(D,q) = 0 \cdot q + 1 \cdot (1-q) = 1-q$, if Player B always employs D. As for the case Player B always uses Strategy C, his expected utility may be: $E(C,q) = 0 \cdot q + 1 \cdot (1-q) = 1-q$. There indeed may be infinite solutions q fulfilling $E(C,q) = E(D,q)$.
[4]It holds, for instance, if $b < a < d < c$. As for the definitions of complementarities and spillovers see, Cooper (1999), for example.

Player B increases the marginal return to the action of Player A. The situation can then give a coordination feature. Define, first of all, a marginal return of action switch. If Player A can strictly switches his strategy from Strategy 1 to Strategy 2, when Player B's strategy adopted Strategy 1, the action switch can give the marginal return:

$$\Delta^{B:1}_{A:1\rightarrow 2} = c - a > 0.$$

If Player B adopted Strategy 2, the same switch may, for instance, give player A the higher marginal return[5]:

$$\Delta^{B:2}_{A:1\rightarrow 2}(A) > \Delta^{B:1}_{A:1\rightarrow 2}(A).$$

Welfare arrangements of choice combination are strictly ordered from higher to lower:

$$(d,d) \succ (a,a).$$

In our *Avatamsaka Game*, there may not found any complementarities relation[6]:

$$\Delta^{B:2}_{A:1\rightarrow 2}(A) = \Delta^{B:1}_{A:1\rightarrow 2}(A).$$

On the other hand, the coordination game has usually another property called *positive spillovers*. This property implies that each player can increase his own reward as a result of the other's switch of action. In our setting, player A can all increase his reward by the other's switch form Strategy 1 to 2.

$$\Delta^{B:2\leftarrow 1}_{A:1}(A) = a - b > 0, \Delta^{B:2\leftarrow 1}_{A:2}(A) = c - d > 0.$$

There exist *positive spillovers* in the Avatansamsaka game[7]:

$$\Delta^{B:1\leftarrow 2}_{A:1}(A) = 1, \Delta^{B:1\leftarrow 2}_{A:2}(A) = 1.$$

Thus it has been seen that there is *no complementarities*, but there are *positive spillovers* in our Avatamsaka game.

---

[5] $d - b > c - a > 0$, since $\Delta^{B:2}_{A:1\rightarrow 2} = d - b$.

[6] $d - b = 1 - 1 = 0; c - a = 0 - 0 = 0$.

[7] $b - a = 1 - 0 = 1; d - c = 1 - 0 = 1$.

11.2 Avatamsaka Game Experiment as a Nonlinear Polya Urn Process

11.2.1 The Elementary Polya Process

By the idea of *evolutionary game*, our game experiment has been designed in the following manner. Conventionally, one exercise may be given the participants of experiment, at first. We usually control the league game of a maximum ten subjects successively for ten times.[8] Every subject can have much information of his rivals. He can always *monitor* a current gain of the other players, as well as their names. *Strategic decisions* always depend the last and the previous results of herself and others reactions. This may mean that the gain distribution will be *path-dependent*.

Now we turn to illustrate an elementary *Polya Urn Process*, which was just introduced into Economics by Brian Arthur, *Snata Fe Institute* and also generalized by him and others,[9] which has been applicable to illustrate *a path-dependent process* as well as to confirm an evolutionary end point like industrial locational by spin-off, and a dual autocatalytic chemical reaction, and the like.[10]

According to Arthur, *Polya Urn Process*, which was formulated by Polya and Eggenbarger (1923), can be illustrated as follows:

> Think of an urn of infinite capacity to which are added balls of two possible colors – red and white, say. Starting with one red and one white ball in the urn, add a ball each time, indefinitely, according to the rule: Choose a ball in the urn at random and replace it: if it is red, add a red; if it is white, add a white. Obviously this process has increments that are path-dependent – at any time the probability that the next ball added is red exactly equals the proportion. We might then ask: does the proportion of red (or white) balls wander indefinitely between zero and one, or does a strong law operate, so that the proportion settles down to a limit, causing a structure to emerge?

Arthur called a limit *proportion* in such a path dependent stochastic process an asymptotic *structure*. As long as the classical proposition is considered, we shall be faced to a limit of proportion, i.e., a simple structure, as Polya proved in 1931. In the classical Polya example, we have a restriction such that the probability of adding a ball of type j exactly equals the proportion of type j.

We apply this kind of Polya stochastic process to our Avatamsaka game experiment to predict an asymptotic structure of the game. In the following we impose the next assumptions for a while.

A 1 We suppose there are a *finite* number of players, $2N$, for instance, who join by each pair into our Avatamsaka game.

---

[8]This experiment demands 45 matches each league game, 450 matches in total, i.e., 45 matches per capita in total.

[9]Yuri M. Ermoliev and Yuri M. Kaniovski, the Glushkov Institute of Cybernetics, Kiev, Ukraina, originally have proved some generalized theorems.

[10]See Chaps. 3, 7 and 10 in Arthur (1994).

A 2 We *identify* the ratio of cooperation, or C-ratio for each player *with* the proportion of the total possible gains for each player.[11]

We just notice that we interpret a red ball as a point of Player A who loves employing Defection Strategy in our two person game, while a white ball as a point of Player B who loves employing Cooperation Strategy, for instance. Player A can then continue to increase his gain if the proportion of his gain is kept higher than Player B. It could be possible, provided that Player B still clinged to his original policy. The Polya original restriction never allow for Player B to change his mind. A *nonlinear* inspection for Players may be taken for account to obtain a more realistic result.[12]

We have usually *ten players* in our game experiment. It may be desirable that the Polya urn may contain *ten colors* in order to analyze *a limit proportion* of the times of defection to cooperation of this game.

11.2.2 A Generalized Polya Urn Process

It is quite interesting to learn that Arthur, a pioneer of *Economics of Complexity* has allocated *one third* of the total pages in his reputable book titled *Increasing Returns and Path Dependence in the Economy*, (Arthur 1994). That famous term like "lock-in" could not be broadly accepted by specialists if a generalization of Ploya process should not be successful. In a generalized version of Polya urn process to allow for a more number of agent(color) than two, a random *vector* X whose components are proportions of each agent, instead of a numerical value(ratio), is to be introduced.

In the line of Arthur, Ermoliev and Kaniovski, we describe *a generalization of path dependent process* in terms of our *repeated game*. Let X_i be a proportion in the total possible points which Player i gains. The initial total *potential* of gains for each player is defined $2N - 1$.[13] In the next period 2, the total maximal gains will grow

[11]This assumption seems much *objectionable*. It however is noteworthy to mention that a tendency towards this relationship *sometimes roughly* holds on the average, as shown in our result of experiment of January 24, 2001 at Chuo University: A set of the average proportion of the total possible gains for each player $= \{0.79, 0.68, 0.60, 0.67, 0.72, 0.74, 0.78, 0.68, 0.73, 0.70\}$, a set of the average C-ratios for each player $= \{0.76, 0.67, 0.53, 0.63, 0.64, 0.76, 0.82, 0.76, 0.73, 0.83\}$. The Spearman's Rank Correlation of the average total gains to the average C-ratios is calculated as 0.627719, while the Kendall's Rank Correlation is 0.535028. Another reference from experiment of December 12, 2000 at Frankfurt University: A set of the average proportion of gains for each player $= \{0.84, 0.70, 0.69, 0.68, 0.82, 0.73, 0.67, 0.72, 0.78, 0.68\}$, a set of the average C-ratios for each player $= \{0.92, 0.89, 0.92, 0.49, 0.8, 0.74, 0.64, 0.51, 0.79, 0.68\}$. The Spearman's Rank Correlation of the average total gains to the average C-ratios is calculated as 0.536585, while the Kendall's Rank Correlation is 0.431818.

[12]Cooperation *by reinforcement* grows herself, if there is a greater ratio of cooperation. Defection can however grow by exploiting a greater population of cooperation.

[13]It is the case if players all mutually cooperate.

by the number of players $2N$. In the period n, therefore, the proportion for Player i at time n will be

$$X_i^n = \frac{b_i^n}{(2N-1)n}.$$

Suppose such *a sample space* that

$$X_i : \Omega \to [0,1],$$
$$\omega \in \Omega \to X_i(\omega) \in [0,1].$$

We can then have a probability $x = X_i(\omega)$ for a sample ω. If a sample implied an experimental result, the sample space Ω could make out *a psychological space* for each agent to give the next move. The probability x may depend on a random variable X_i. If we have $2N$ agents, it then follows a random vector of proportions X^n of period n:

$$X^n = \left(X_1^n, X_2^n, \ldots, X_{2N}^n\right).$$

where it is assumed that each element X_i^n mutually is *independent*, in other words, each agent is *psychologically independent*. Players start at a vector of the initial gains distribution

$$b^1 = \left(b_1^1, b_2^1, \ldots, b_{2N}^1\right).$$

Let $q_i^n(x)$ define a probability of player i to *earn* a point *by means of Strategy C* in period n[14]. Thus a series of proportions

$$\{X^1, X^2, X^3, \ldots, X^n\}$$

will be generated after the iteration of n times.

In the following, q^n behaves as a rule for *a mapping* from X^n to X^{n+1}. Consider a one-dimensional dynamics.

$$q_i(X_i)(\omega) \in [0,1],$$
$$q_i(x) : \Omega \to [0,1].$$

If X^n, as a random variable giving a proportion x, appeared at a low level, a value of X^{n+1} might be greater than the previous value of X^n. This case is implied by *A2*

[14]Speaking strictly, probability q may be regarded as an *ex post* probability in a sense of Bayesian.

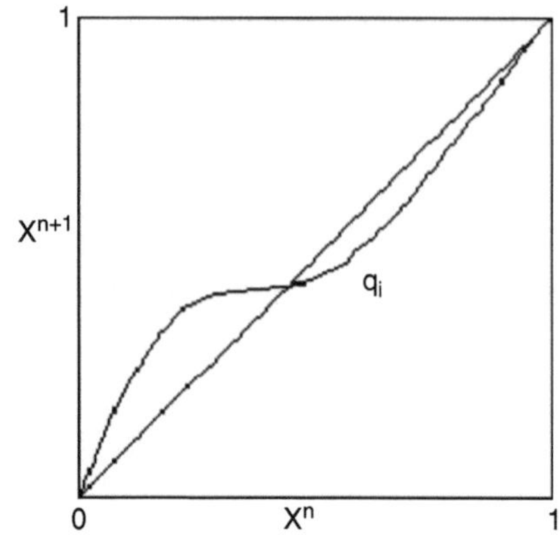

Fig. 11.2 A nonlinear function form of q

that a low level of proportion in gain can induce an increase of the ratio of cooperation in an individual strategy.[15] If q is *nonlinear*, a *nonlinear* dynamics may be generated. See Fig. 11.2.

11.2.3 A Nonlinear Polya Process

We have already defined a function q for a $x = X_i(\omega)$, on one hand. Now, on the other hand, we define a function β for the same ω:

$$\beta_i(X_i)(\omega) \in \{0, 1\},$$
$$\beta_i(x) : \Omega \to \{0, 1\}.$$

It then follows, if $x = X_i^n(\omega)$,

$$\beta_i^n(x) = 1 \text{ with pobability } \textit{given} \text{ by } q_i^n(x),$$
$$\beta_i^n(x) = 0 \text{ with probability } \textit{given} \text{ by } 1 - q_i^n(x).$$

Each player might have a maximum gain of $2N - 1$ from the $2N - 1$ matches. Hence the *dynamics* for addition of point by Player i may be written:

$$b_i^{n+1} = b_i^n + (2N - 1)\beta_i^n(x).$$

b_i^n then shows an *accumulated* gain at time n through the iteration of game for Player i. Thus the *nonlinear evolution* of the proportion for Player i will, by substitution of X^n, be given in the form of[16]

[15]In a certain circumstance, it may be true.

[16]Dividing the both sides by $(2N - 1)(n + 1)$ and also multiplying them $\frac{(2N-1)(n+1)}{(2N-1)n}$, it follows the difference equation.

$$X_i^{n+1} = X_i^n + \frac{1}{n+1}\left(\beta_i^n(x) - X_i^n\right).$$

Reformulating this equation in view of

$$\xi_i^n(x) = \frac{1}{n+1}\left(\beta_i^n(x) - q_i^n(x)\right),$$

we obtain

$$X_i^{n+1} = X_i^n + \frac{1}{n+1}\left(q_i^n(x) - X_i^n\right) + \xi_i^n(x).$$

Notice that *the conditional expectation* of ξ_i^n is zero.[17] Thus we have reached *the expected motion*:

$$E(X_i^{n+1}|X^n) = X_i^n + \frac{1}{n+1}\left(q_i^n(x) - X_i^n\right),$$

from which we can easily verify that motion tends to be directed by the term $q_i^n(x) - X_i^n$. Arthur and others call this process *a nonlinear Polya process* when the form of q is *nonlinear*, as indicated in Fig. 11.2 in the above.

The above process is just *a martingale*. By a direct application of the theorems of Arthur, Ermoliev and Kaniovski, we could justify a convergence of the C-ratios in the community *not* to the extreme points like 0, 1.

Players shall initially be motivated by the behaviors of the other players. Eventually, however, players' behavior could be independent from the others.

References

Arthur WB (1994) Increasing returns and path dependence in the economy. University of Michigan Press, Ann Arbor

Aruka Y (2000) Avatamsaka game structure and it's design of experiment. Working Paper No.3, IBRCU, Chuo University

Aruka Y (2001) Avatamsaka game structure and experiment on the web. In: Aruka Y (ed) Evolutionary controversies in economics. Springer, Tokyo, pp 115–132

Cooper RW (1999) Coordination games. Cambridge University Press, Cambridge

Kamata S (1988) Kegon no Shiso (The Thought of Avatamsaka). Kodan-sha, Tokyo (in Japanese)

Polya G (1931) Sur quelques Points de la Theorie des Probabilites, Ann Inst H Poincare 1:117–161

Polya G, Eggenberger F (1923) Ueber die Statistik verketteter. Z Angew Math Mech 3:279–289

[17]By the definition of β, it follows that $E(\xi_i(x)) = 0.5\frac{1}{n+1}(1-q_i(x)) + 0.5\frac{1}{n+1}(0-(1-q_i(x)) = 0$.

Chapter 12
Non-Self-Averaging of a Two-Person Game with Only Positive Spillover: A New Formulation of Avatamsaka's Dilemma

Yuji Aruka and Eizo Akiyama

In this game (Aruka 2001), selfishness may not be determined even if an agent selfishly adopts the strategy of defection. Individual selfishness can only be realized if the other agent cooperates, therefore gain from defection can never be assured by defection alone. The sanction by defection as a reaction of the rival agent cannot necessarily reduce the selfishness of the rival. In this game, explicit direct reciprocity cannot be guaranteed. Now we introduce different spillovers or payoff matrices, so that each agent may then be faced with a different payoff matrix. A ball in the urn is interpreted as the number of cooperators, and the urn as a payoff matrix. We apply Ewens' sampling formula to our urn process in this game theoretic environment. In this case, there is a similar result as in the classic case, because there is "self-averaging" for the variances of the number who cooperate. Applying Pitman's sampling formula to the urn process, the invariance of the random partition vectors under the properties of exchangeability and size-biased permutation does not hold in general. Pitman's sampling formula depends on the two-parameter Poisson–Dirichlet distribution whose special case is just Ewens' formula. In the Ewens setting, only one probability α of a new entry matters. On the other hand, there is an additional probability θ of an unknown entry, as will be argued in the Pitman formula. More concretely, we will investigate the effects of different payoff sizes from playing a series of different games for newly emerging agents. As Aoki and Yoshikawa (2007) and Aoki (2008) dealt with a product innovation and a process innovation, they criticized Lucas' representative method and the idea that players face micro shocks drawn from the same unchanged probability distribution. In the light of Aoki and Yoshikawa (2007), we show the same argument in our Avatamsaka game with different payoffs. In this setting,

Y. Aruka (✉)
Faculty of Commerce, Chuo University, Higashinakano, Hachioji, Tokyo 192–0393, Japan
e-mail: aruka@tamacc.chuo-u.ac.jp

E. Akiyama
Graduate School of Systems and Information Engineering, University of Tsukuba, Tennodai 1-1-1, Tsukuba, Ibaraki 305-8573, Japan
e-mail: eizo@sk.tsukuba.ac.jp

Y. Aruka (ed.), *Complexities of Production and Interacting Human Behaviour*,
DOI 10.1007/978-3-7908-2618-0_12,

innovations occurring in urns may be regarded as increases of the number of cooperators in urns whose payoffs are different.

12.1 Introduction

12.1.1 Towards a New Method to Deal with Two Conspicuous Features of Modern Society

When discussing modern society, there are two conspicuous features: a huge number of players and a set of unexpected results.

First, it is trivial that there are a huge number of players with their intensive and/or extensive interactions. The total population of the world is 6.5×10^9 players, and is gradually approaching $\times 10^{10}$. Including global financial activities, the order of magnitude of interactions will be 10^{15} or so. The interaction of human activities might amount to interactions based on the number of molecules in a box of gas, i.e., 10^{23}.[1] Social sciences must now treat a huge number, just as in statistical physics.

Second, in this secular world, people with good intentions are not always rewarded. This perception could actually be traced back to ancient beliefs such as Tyche or Fortuna of ancient mythology. In the social sciences, this one belief has inspired and promoted applications of probability theory since David Hume.

Briefly look at this story in light of Mainzer's work (Mainzer 2007a, b; Aruka 2009). In the ancient cultures, the goddesses Tyche in Greece and Fortuna in Rome were well known. Fortuna however does not always guarantee fortunes for individuals. According to Mainzer (2007b), the Christian tradition rejected the idea of goddesses of fortune and fate, and Kairos, the opportune moment, instead became highlighted in the new era. Individuals can prepare for the moment that may bring them opportunities. Mainzer (2007a), in connection with nonlinear scientific innovations of the last century, insightfully argued Kairos as a "right moment" in order to judge Mandelbrot's parables (Mandelbrot and Hudson 2004): the Joseph effect and the Noah effect. In the modern world with its increasing complexity, it is relevant to describe contingencies in connection with Kairos, a kind of a macroscopically weak control mechanism.

Reconsidering "a free act of human self-determination in a stream of nonlinearity and randomness in history," the classical philosophical arguments could be revived. It is easy to realize that individuals never rely only upon their own a priori preferences. On the other hand, "chances" cannot be ignored independently of "individual abilities" (including learning abilities) and "circumstances". Things like novelties emerging either from nature or society could never happen without intermediation by chance. Thus, how chances may intermediate emergences of thinking must be explored, so that we are ready to argue Hegel's methodology positively in the new methodological

[1]See Fujiwara (2008).

view. Then it is certain that "Good intentions may lead to bad effects without their subjective intentions. Hegel called it a stratagem of reason" (Mainzer 2007a, p. 408).

In the following discussions, equilibrium does not imply a kind of "self-filling prophecy" game result for participating agents, whose expectations must often be circumvented.

12.1.2 An Interactive Chance

Kant distinguished contingencies as being more empirical, more logical, and more intelligible. An empirical contingency that depends on a certain cause is meant as the "event". Sun irradiation is not a necessary condition for heating a stone, but a stone necessarily falls – on the basis of gravity – to the earth at any moment. Seemingly, these are opposite observations. A concept may fail to catch all the events around it. A property (attribute or type) that does not follow from the definition of a concept (category) is called logically "chance" (Mainzer 2007a, p. 408).

This is the reason why the idea of type distribution with exchangeable agents should be used, because it is difficult to identify the attributes or types by themselves with reference to events at hand. Due to the limitations of the empirical data available, however, a microscopic approach in terms of the given or fixed types must be given up, using some idea of state property instead. An example is income transition in the society. In this example, middle class property is a state property, if the income range of the middle class is pertinently defined. The ingredients of various types in the middle class of income is then ignored. Such a simplification is called a metonymy from types to state properties, according to Hildenbrand (1994). Exchangeable agents are virtually used, instead of types in a precise sense. Types are replaced with state variables. A state variable then becomes a surrogate variable for type. Thus, the focus is on the combinatorial dynamics of the total number of state and the state variables. It may be helpful to reproduce Fig. 12.1[2] in order to deal with transitional and exchangeable types.[3]

12.1.3 A Macro–Microscopic Linkage and Our Approach

The "Price Equation" (Price 1970, 1972) in biology is well known as a simple statistical statement on the expected change of the frequency of a gene. Recently,

[2]This figure is almost equivalent to Figure 3 in Aruka (2007).

[3]This interpretation is supplemented by citing Bowles (2004), in terms of behavioral dynamics: "Thus individuals are the bearers of behavioral rules. Analytical attention is focused on the success or failure of these behavioral rules themselves as they either diffuse and become pervasive in a population or fail to do so and are confined to minor ecological niches or are eliminated. The dramatis personae of the social dynamic thus are not individuals but behavioral rules: how they fare is the key; what individuals do is important for how this contributes to the success or failure of behavioral rules".

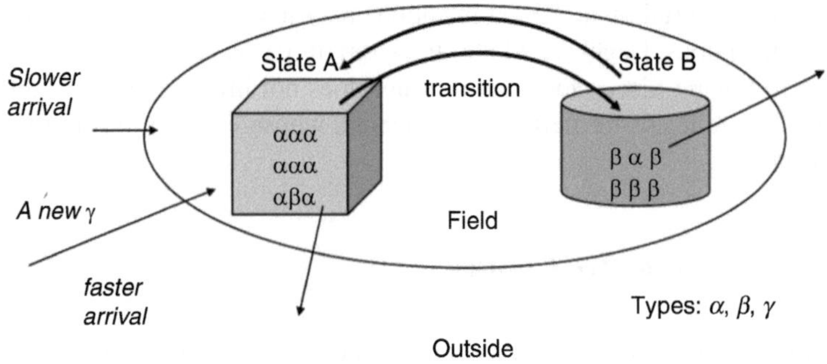

Fig. 12.1 The idea of transmission mechanism

the *Journal of Economic Behavior and Organization* (vol. 53(1) 2004) has arranged a special issue on the Price Equation in the context of Henrich's contributions (Henrich 2004). In this gene, x_i denotes an altruistic allele if individual i retains $x_i = 1$, while an egoistic allele if $x_i = 0$. x_i can then express the current frequency of this gene. Let w_i be the number of offspring of i, namely, the absolute fitness of i while $\bar{w}_i$ is the average fitness of the population. In Henrich's contribution (2004), "we ignore any effects arising from the transmission process (e.g., recombination, mutation, etc.)". It then follows:

$$\bar{w}\Delta\bar{x} = \beta_{w_j x_j} Var(x_j) + E(\beta_{w_{ij} x_{ij}} Var(x_{ij})) \tag{12.1}$$

Selection between-groups + Selection within-groups

Here $\beta_{w_{ij} x_{ij}}$ means the within-group regression coefficients of x_{ij} with respect to w_{ij} in group j. This type of Price Equation "tells us that the change in the frequency of allele created by natural selection acting on individuals can be partitioned into between-group and within-group components" (Henrich 2004).

This is a good practice to measure contributions of interactive factors due to between-groups and within-groups. It is not a better one, however, because this description lacks an overarching mechanism working between within-groups and between-groups. In the above equation, this is just a transition mechanism in the Master Equation approach. Without this overarching mechanism, the signs of between-groups effects as indicative of group selection may not be definitely ascertained (Aruka 2004).

The assumption of a mutant gene to promote altruistic behavior for this group may be rejected because it reduces the fitness of its carrier. In order to make group selection feasible, a transmission mechanism is indispensable, as Henrich stated (2004). Cultural group selection based on such a transmission could be viable if and only if a certain macroscopic order is implemented in the interactive field. In other words, a promising way to take a cooperative system formation into account is to introduce a macroscopic order in the concerned field in which individual agents act bilaterally with groups. If such a binding condition of group selection to form a macroscopic order could be found, an overarching mechanism could be found to illustrate the processes of a dispersive social system.[4]

Thus, even in the case of describing such an interactive process between cooperation and defection, we could be induced to adopt the stochastic approach employed by Masanao Aoki's seminal works (Aoki 1996, 2002, 2008; Aoki and Yoshikawa 2006, 2007), written partly in cooperation with Hiroshi Yoshikawa. It might be hypothetically true that human beings have both the altruistic gene and the egoistic gene. However, individuals never rely only upon their own distinct genes, since chance must be consecutively created in the course of the sequential interactions of variant clusters of exchangeable agents. So, interactive chances matter.

12.2 A Two-Person Game with Positive Spillover

12.2.1 A Famous Tale from Buddhism Sutra

There is a famous tale of *Avatamsaka Sutra*, mahayana buddhist sutra. Suppose that two men sat down at a table, across from one another. They are tied up with rope except for one arm, then each is given a long spoon. They cannot feed themselves because the spoon is too long, but each can feed the other. There is plenty of food on the table.

- If they cooperate to feed each other, they can both be happy. This defines Paradise.
- One may be so kind as to feed the other but the other may not cooperate. This must give rise to a feeling of hate for the other. This describes a situation of Hell.

The gain structure will not essentially depend on the altruistic willingness to cooperate. This tale often is cited all over the world, recently, in *Chicken Soup for the Soul* (Canfield and Hansen 2001), a best-selling book in the U.S.

Aruka (2001) formulated this tale as two-person, two-strategy game and called this game Avatamsaka game. The main features of the Avatamsaka game are summarized as:

- Any defector may use his rival's cooperation to get his gain. Without the rival's cooperation, however, his gain cannot be guaranteed.
- The expected value of a gain for any agent could be reinforced, if it should be generated by a higher average rate of cooperation as measured by the frequency of mutual cooperation. Here there is a macroscopically weak control mechanism, which does not mean a personal mutual fate control described in Thibaut and Kelley (1959).

[4]Hildenbrand (1994) called it "macroscopic microeconomic" linkage.

- The sanction by defection as a reaction of the rival agent cannot necessarily reduce the selfishness of the rival. In this game, then, any explicit direct reciprocity cannot be automatically guaranteed.

In the Avatamsaka game (Aruka 2001), selfishness may not be determined even if the agent selfishly adopts the strategy of defection. Individual selfishness can only be realized if the other agent cooperates. Any certain gain from defection can never be assured by defection alone. The sanction by defection as a reaction of the rival agent cannot necessarily reduce the selfishness of the rival. In this game, any explicit direct reciprocity cannot be guaranteed.

12.2.2 A Historical Note

12.2.2.1 The "Mutual Fate Control" Game

The authors recently learned that the same formulation, looking at the payoff structure formally, had already been explored in the field of psychology. In psychology, it is very important to know how one could affect the other's behavior. Our payoff structure can show this. In the field of social psychology, the payoff structure of the Avatamsaka game was argued. The same game appeared earlier, in 1959, in the book titled *The social psychology of groups* (Thibaut and Kelley 1959). The game was called "mutual fate control". Psychologists created experiments based on this game. Recently, Mitropulos revived it (Mitropoulos 2004).

12.2.2.2 The Path Dependence of the Avatamsaka Game

Our original view focused on an emerging/evolving environment, i.e., path dependency. We further focus on two kinds of averaging.

- Self-averaging: eventually players' behavior could be independent from the other players.
- Non-self-averaging: the invariance of the random partition vectors under the properties of exchangeability and size-biased permutation does not hold in general.

12.2.3 Geometric Structures of Two-Person Games

In order to clarify the characteristics of the Avatamsaka game, we compare our Avatamsaka game with a class of Prisoner's Dilemmas, taking two numerical examples shown in Tables 12.1 and 12.2.

Table 12.1 Avatamsaka

	Player 2	
Player 1	Cooperation	Defection
Cooperation	(1, 1)	(0.25, 1)
Defection	(1, 0.25)	(0.25, 0.25)

Table 12.2 Prisoner's dilemma

	Player 2	
Player 1	Cooperation	Defection
Cooperation	(0.7, 0.7)	(0.1, 0.9)
Defection	(0.9, 0.1)	(0.3, 0.3)

$$\text{Avatamsaka}\begin{bmatrix} R & S \\ T & P \end{bmatrix} = \begin{bmatrix} 1 & 0.25 \\ 1 & 0.25 \end{bmatrix}$$

$$\text{PD}\begin{bmatrix} R & S \\ T & P \end{bmatrix} = \begin{bmatrix} 0.7 & 0.3 \\ 0.9 & 0.1 \end{bmatrix}$$

In order to clarify the properties of the two games, we may use the letters R, S, T, P and the differences of $D_g = T - R, D_r = P - S$.

Here we call

$$D_r = P - S$$

the "Risk Aversion Dilemma" and

$$D_g = T - R$$

the "Risk Seeking Dilemma".

First, the Avatamsaka game is specified as follows:

$$D_g = T - R = 0\,, \tag{12.2}$$

$$D_r = P - S = 0\,. \tag{12.3}$$

On the other hand, a Prisoner's Dilemma game is specified as follows:

$$D_g = T - R = 0.2\,, \tag{12.4}$$

$$D_r = P - S = 0.2\,. \tag{12.5}$$

Note that dilemmas may be generated unless "complementarity" for both players holds.

We call

$$R - S = T - P$$

the "spillover". In the Avatamsaka game, spillovers are always positive. So each player can then increase his rewards by the other player's strategy switching (Aruka 2001, p. 118). [Regarding the Avatamsaka original papers, see Aruka (2001, 2002) and Akiyama and Aruka (2006)].

12.2.4 Tanimoto's Geometrical Expression of a Two-Person Game

Tanimoto geometry (Tanimoto 2007a, b; Tanimoto and Sagara 2007) is used to describe the characteristics of games.

The Tanimoto geometry is composed by the following equations:

$$P = 1 - 0.5r_1 \cos\left(\frac{\pi}{4}\right) \tag{12.6}$$

$$R = 1 + 0.5r_1 \cos\left(\frac{\pi}{4}\right) \tag{12.7}$$

$$S = 1 + rr_1 \cos\left(\frac{\pi}{4} + \theta\right) \tag{12.8}$$

$$T = 1 + rr_1 \sin\left(\frac{\pi}{4} + \theta\right) \tag{12.9}$$

$$\text{Here } r = \frac{r_2}{r_1}; r_1 = PS, r_2 = SM\,.$$

Using this description, we can illustrate the key points R, S, T, P on the two-dimensional plane of Fig. 12.2. Here the horizontal axis shows Player 1's payoff while the vertical axis shows Player 2's. An Avatamsaka game is shown in Fig. 12.3 and a Prisoner's Dilemma game is shown in Fig. 12.4.

We define the contour on the parametric plane (θ, r), depicting the contour of $T + P = R + S$ on this plane. A contour is shown in Fig. 12.5; $T + P > R + S$ is inside the contour and $T + P < R + S$ is outside the contour. We show these relationships in Fig. 12.6. A Prisoner's Dilemma game is to be located on the vertical line $\theta = \frac{\pi}{2}$. From this point of view, a Prisoner's Dilemma game may be regarded as a reference standard for any two-person game.

Note that our Avatamsaka game has a property of $T + P = R + S$. This implies that our game could always be defined as the contours, if the various payoffs were allowed to be introduced, as we observe later. Different contours correspond to different payoffs of the Avatamsaka game, so players can walk around between different states of payoffs. In the latter case, our argument may be ubiquitous on the contours plane. The Avatamsaka game by definition has infinite equilibria even given a fixed payoff structure. Originally, our game was always allowed to be on

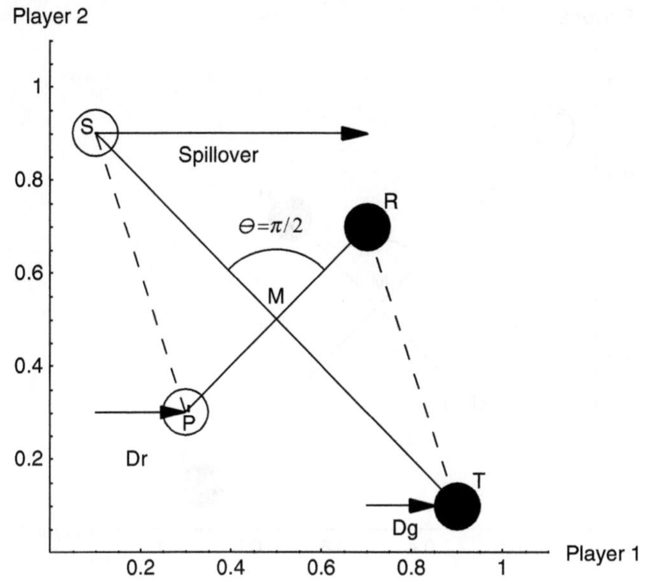

Fig. 12.2 The Tanimoto geometry

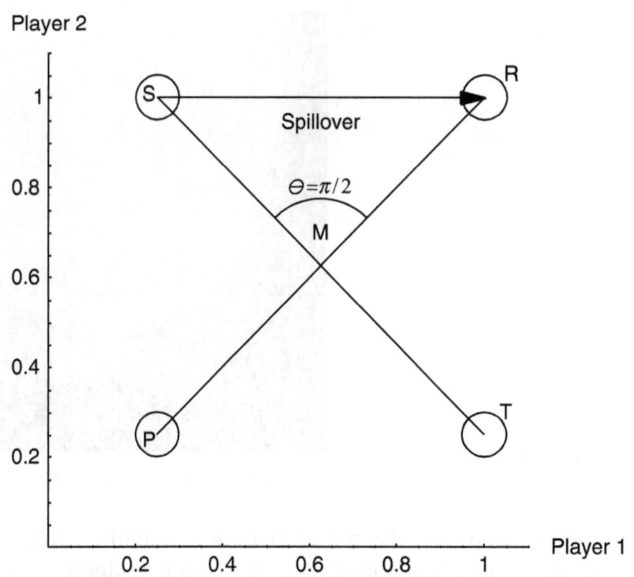

Fig. 12.3 Avatamsaka

any Nash equilibrium. The dynamics of our game are a kind of equilibrium dynamics. In the following stochastic setting, by resorting to Aoki's method, players could never have their reasonable self-filling prophecies even if players

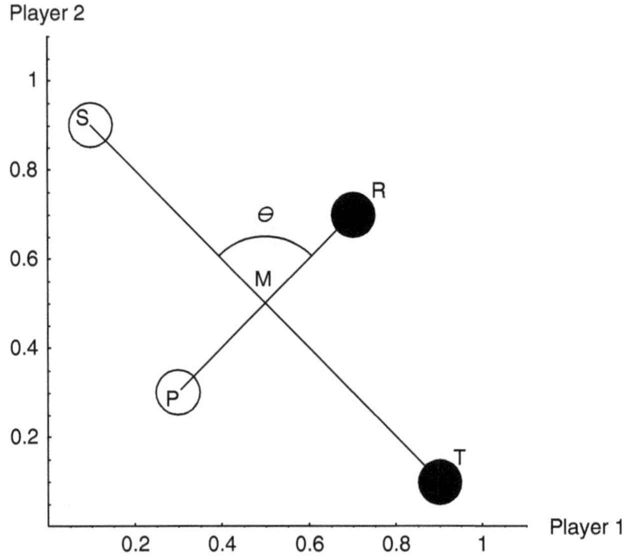

Fig. 12.4 Prisoner's dilemma

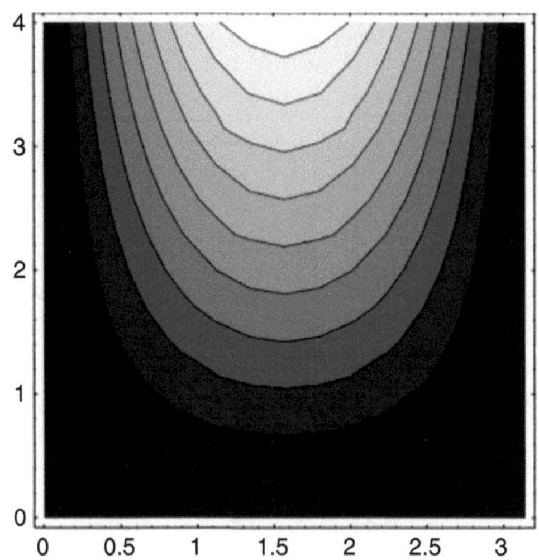

Fig. 12.5 The contour of T + P = R + S

should wander on the contours. Also notice that our notion of equilibrium does not necessarily guarantee each agent's prophecy. This is a new finding.

Finally, compare the Avatamsaka game with the donor–recipient type prisoner's dilemma, which can be regarded as an exceptional case ($c = 0$) of ($T = b, R = b - c$, $P = 0$ and $S = -c$).[5] In our game, such a cost is not observed, instead of taking

[5]This was pointed out by an anonymous referee to whom we are indebted for this description.

Fig. 12.6 The relationships on the contours

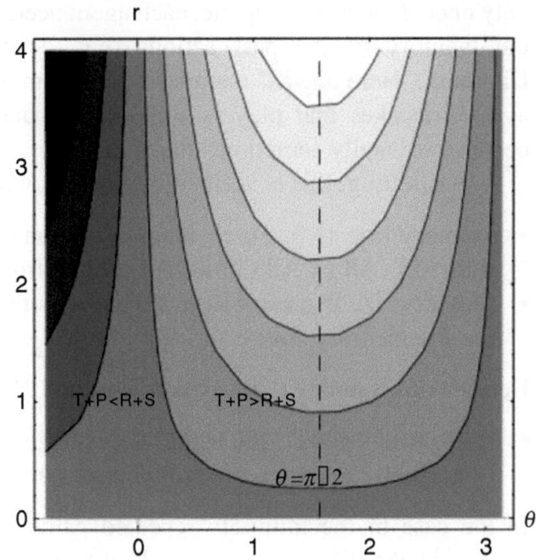

account of transition risks posed by the failure of agent expectations.[6] In this article, the dynamics of expectation – which are not based on the benefit–cost relationship – matter. Our game could be the suggested case by linear transformation if a small perturbation is given on the off-diagonal elements of the payoff structure. An introduction of perturbation suggests that our game becomes out of equilibrium. Such disequilibrium must be important but, for the moment, we confine ourselves to observing only equilibrium dynamics.

12.2.5 A Last Attempt to Formulate the Avatamsaka Game in Terms of an Evolutionary Game

In Akiyama and Aruka (2006), the Avatamsaka game as an evolutionary game of dyadic pairs of interactive agents was discussed. We focused on the so-called memory effects on agent decisions, supposing a repeated game with "action noise" (or matching noise).[7] In reality, interaction with a person does not happen

[6]The risk may be the risk of realized expectation, whether the expectation is cooperation or defection, when agents wander around.

[7]As Bowles (2004, p. 62) argued, this technique is not superficial: "What is called matching noise is another way that chance affects evolutionary dynamics. When small numbers of individuals in a heterogeneous population are randomly paired to interact, the realized distribution of types with whom one is paired over a given period may diverge significantly from the expected distribution. The difference between the realized distribution and the expected distribution reflects matching noise and may have substantial effects. . . . How, then, are evolutionary models different? First, mutations, behavioral innovations, and matching noise are distinct because these sources of stochastic events are endogenous to evolutionary models".

only once. In a repeated game, each agent needs a strategy to decide her action from the memory of the past actions (e.g., "Tit-For-Tat" for repeated Prisoner's Dilemma). Here agents' memory size becomes important. Action noise is defined as the mistakes that players sometimes make when they hand in their actions because we really sometimes make mistakes.

In our setting, this is verified by the following propositions:

- Memory size is 0: There is no evolutionarily stable strategy. The population shares of (All *D*, All *C*)[8] do not change from the initial generation.
- Memory size is greater than 2: Pavlov is the *ESS* (evolutionary stable strategy) for any memory size.

Here Pavlov's policy (Nowak and Sigmund 1993) is defined in the following way:

- If she feels "good", she would not change her current action.
- If she feels "uneasy", she will change her current action.

In the case of the infinitely repeated game, we have a definite conclusion that Pavlov is the *ESS* for $m \leq 2$. If the probability of another repetition of the Avatamsaka game is less than 1.0, Pavlov is the *ESS* for any memory size.

This result can easily be applied to any dependent game like the Avatamsaka game (with any number of players and with any number of strategies). Thus Pavlovian strategy is the *ESS* in a dependent game: *Keep changing my action until the others benefit me.*

Now we will try to formulate the Avatamsaka game in view of stochastic processes where we can deal with a truly emerging mutant. In contrast with our last results, stable equilibrium states may not necessarily be guaranteed.

12.3 The Avatamsaka Stochastic Process Under a Given Payoff Matrix

12.3.1 The Simplest Polya's Urn Process

In the original stage, an urn contains a white ball and a red ball only. Draw out one ball, and return that ball with an additional ball of the same color. Repeat this trial again and again. The number of balls increases by one with each trial. So after the completion of two trials,

- The total number of balls after the second trial $= 2 + 1 + 1 = 4$.
- Hence the expected value of white balls after two successive trials is $2(= 4/2)$ since the number of red ball is equal to the number of white ball at the initial point.

[8]*D* denotes "Defection" while *C* denotes "Cooperation".

After the completion of n trials, what is the probability that the urn contains just one white ball? This must be equivalent to the probability of n-fold successive draws of red balls.

$$\frac{1}{2}\frac{1+1}{2+1}\frac{1+1+1}{2+1+1}\cdots\frac{n}{n+1}=\frac{1}{n+1}.$$

path - dependent

Suppose the total number of white balls can be k at the l-th trial $(1 \le k \le n+1)$. We describe this situation as

$$P(l,k)=\frac{1}{l+1}.$$

By induction, it is easy to prove that

$$P(n,k)=\frac{1}{n+1}.$$

This result shows that any event (n, k) at the n-th trial can emerge, i.e., any number of white balls can go everywhere at the ratio of $\frac{1}{n+1}$.

A short demonstration of the classic, simplest case of the Polya urn process (Figs. 12.7 and 12.8):

In this kind of short simulation, it is trivial that the trend of this trial must be seen within a short time like 100 *trials.*

12.3.2 Avatamsaka Characteristics

Avatamsaka characteristics are summarized as follows:

- The increase of the ratio $\frac{C}{C+D}$ (C-ratio) implies an average increase of C agents who changed their strategies from D to C, which leads to the expectation of mutual cooperation. This reflects the property of positive spillover.

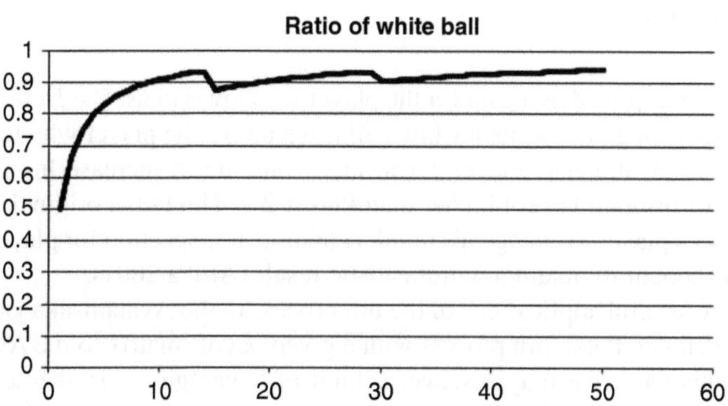

Fig. 12.7 Simulation 1

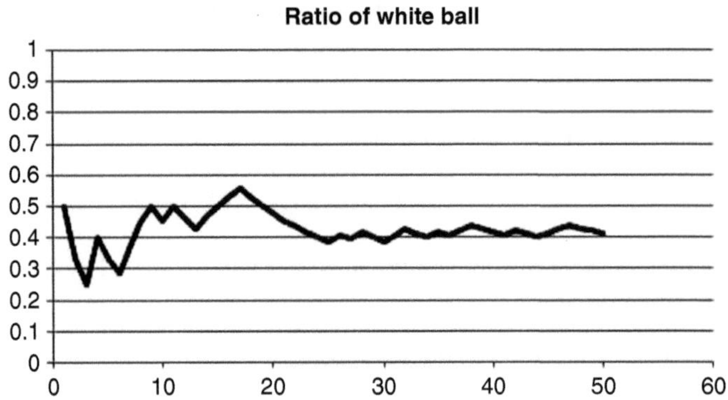

Fig. 12.8 Simulation 2

- In the simple case, then, it is trivial that the expectation of mutual cooperation could be reinforced.
- So this game has a macro structure of cooperation that can loosely control a path of interaction.

Now, apply a Polya urn stochastic process to our Avatamsaka game experiment to predict an asymptotic structure of the game, according to Aruka (2002). This stochastic process was employed by Brian Arthur (1994) and in particular in Arthur et al. (1987). In order to apply Arthur's theorem, we impose the assumptions:

Assumption 1 There are a finite number of players who join the Avatamsaka Game by pairs.
Assumption 2 The ratio of cooperation, or C-ratio for each player, is identified with the proportion of the total possible gains for each player.

12.3.3 New Applications of the Urn Process to Avatamsaka Game

A red ball is interpreted as a gain for the player who loves to use the Defect Strategy in our two-person game, while a white ball is a gain for the player who loves to use the Cooperation Strategy. Player 1 can then continue to increase his gain if the proportion of his gain is kept higher than Player 2's. The Polya original restriction never allows a player to change his mind. A nonlinear inspection for players may be taken into account to obtain a more realistic result (Aruka 2001).

There are several applications of the urn process to the Avatamsaka game. Here, we apply a classic Polya urn process with a given payoff matrix to the Avatamsaka game. In this case, we may discover a kind of "averaging", in the sense that a player's behavior could be independent from the others. Fortunately, however, we have more candidates for doing this. We have the next two sampling formulas.

- Ewens' sampling formula: a K-dimensional Polya urn process with multiple payoff matrices and possible new agents.
- Pitman's sampling formula: a two-parameter Poisson–Dirichlet distribution.

In the latter case, there may be a "non-averaging" property in the sense that the invariance of the random partition vectors under the properties of exchangeability and size-biased permutation does not hold in general (Pitman 2002).

12.3.3.1 A Comparison with the Market

In the stock or commodity exchanges, any trade or sequence of trades must be settled at the end of session. Once started, any trade must have an end within a definite period, even in the futures market. A distribution of trader's types can affect the trading results, but not vice versa. Compared with the stock exchange system, each game round in a repeated game must change its own environment. A distribution of agent's types can affect the results of the game, and vice versa. Agents must inherit their previous results. This situation describes path-dependency in a repeated game.

12.3.4 Self-Averaging in the Avatamsaka Game with a Special Rule

12.3.4.1 A Balanced Urn Game

According to Aoki and Yoshikawa (2007), there is another urn model to discuss: the idea of a "balanced urn" in the context of Flajolet et al. (2005).

The urn contains two different colored balls: white and red. If a white ball is drawn from the urn, a white balls and $b - a$ red balls are returned into the urn. If a red ball is drawn from the urn, b red balls are returned into the urn. This type of ball replacement is expressed in the replacement matrix, R:

$$R = \begin{pmatrix} a & b-a \\ 0 & b \end{pmatrix}.$$

The urn is called balanced, if

- The sum of the first row $= a + b - a = b$,
- The sum of the second row $= b$.

The matrix R then is a balanced triangular urn.

Proposition 1. *The number of white balls in the balanced triangular urn model is non-self-averaging.* (For the proof, see Aoki and Yoshikawa 2007, pp. 11–13). □

Before discussing the properties of non-self-averaging of strategy arrangements in a more general case, we apply this modeling directly to our particular Avatamsaka game: if a player is faced with the cooperation of a rival, the player can react by C-ratio of $\frac{a}{b-a}$. On the other hand, the player can react merely by defecting when faced with the defection of a rival.[9] The game must be non-self-averaging according to Proposition 1.[10]

In the balanced urn, the total number of balls in the urn is constant regardless of the color of a drawn ball. The mixture of the two colors (types) must be path-dependent. This may create a non-self-averaging property for the game. If a player is faced with the cooperation of a rival, the player can react by the C-ratio of a to $b-a$. On the other hand, the player can react merely by defecting when he is faced with the defection of a rival; b is constant as the number of defections is always constant. This special unchanged attitude of D-agents can rather contribute a stable expectation of non-self-averaging.

12.3.4.2 A Nonlinear Polya Urn Dynamic Game: A Gain-Ratio in the Total Possible Points

Let X_i be a gain-ratio in the total possible gains that Player i gains. The gain-ratio $= \frac{\text{Gain}}{\text{Total Gain}}$. The initial total potential of gains for each player is defined as $2N-1$. In the next period, 2, the total maximal gains will grow by the number of players $2N$. In the period n, therefore, the gain-ratio for Player i at time n will be

$$X_i^n = \frac{b_i^n}{(2N-1)n}.$$

Suppose there is such a sample space Ω that

$$\begin{cases} X_i : \Omega \to [0,1] \\ \omega \in \Omega \to X_i(\omega) \in [0,1]. \end{cases}$$

There is a probability $x = X_i(\omega)$ for a sample ω. Then, the nonlinear evolution of the proportion, i.e., Arthur's dynamics:

$$X_i^{n+1} = X_i^n + \frac{1}{n+1}\left(\beta_i^n(x) - X_i^n\right). \tag{12.10}$$

Inspecting the expected value of X_i^n leads to the result:

---

[9]This statement needs an assumption like Assumption 3, given in Section 4.

[10]We can add to a more general proposition: in non-balanced triangular urn models as depending on the values of parameters, no-self-averaging emerges (Aoki and Yoshikawa 2007, p. 14).

Proposition 2. *Players shall initially be motivated by the behaviors of the other players. Eventually, however, players' behavior could be independent from the others.* (For the proof, see Aruka 2002) □.

12.3.5 Relaxing the Assumption of a Given Payoff Matrix

Suppose there are a number of different urns with various payoff matrices, and each has its own size of spillover. A different spillover in our Avatamsaka game may change the inclinations of players' reactions, but these inclinations are not necessarily symmetrical.

12.3.5.1 Inclinations of a Player's Strategy

An urn with a greater spillover might sometimes be more attractive for a cooperative player in a game with a high level of cooperation, because the players could earn greater gains. A cooperative player in a game with a low level of cooperation might be attracted to enter an urn with a smaller spillover than in a game with a much higher level of cooperation. But, he can become a defecting player. There may thus be various players' plans for their strategies. Any player depends on his state, which may change, while an urn, that is, a payoff matrix occurs stochastically.

12.3.5.2 Nothing to Do with Any Particular Transition Mechanism

Thus, as we suppose that any urn must be random for players, any player must be random for urns. In our Avatamsaka game, the size effects due to different spillovers are irrelevant, so any specific or particular transition mechanism need not be assumed ex ante.

12.4 The Key Ideas for the New Economics

12.4.1 The Economics of the Master Equation and the Fluctuations

The stochastic evolution of the state vector can be described in terms of the master equation as equivalent to the Chapman–Kolmogorov differential equation system. The master equation leads to bringing the aggregate dynamics, from which the Fokker–Planck equation could be derived. Thus, we can explicitly argue the fluctuations in a dynamic system. These settings can indispensably be connected with the following key ideas, making the type classification of agents in the system feasible, and tracking the variations in cluster size, as featured by statistical physics and combinatorial stochastic processes (Aoki and Yoshikawa 2006).

In Aoki's new economics, there are exchangeable agents in the combinatorial stochastic process, as in the urn process. The exchangeable agents come out by the use of random partition vectors, as in statistical physics or population genetics. The partition vector provides the state information, allowing discussion of the size-distribution of the components and their cluster dynamics with the exchangeable agents.

Define a maximum countable set, in which the probability density of transition from state i to state j is given, respectively. In this setting, the dynamics of the heterogeneous interacting agents gives the field where an agent can become another agent. It is also important to note that this way of thinking welcomes the unknown agents.

12.4.1.1 A K-Dimensional Polya Distribution

A K-dimensional Polya distribution is stated by the use of parameter θ. We then have a transition rate:

$$w(\mathbf{n}, \mathbf{n} - \mathbf{e}_i + \mathbf{e}_j) = \frac{n_i}{n} \frac{n_j + \theta_j}{n - 1 + \theta}$$

where

$$e_i = (0, \ldots, 0, 1, 0, \ldots, 0).$$

1 is located i-th position in the above row. It is also noted that

$$n = n_1 + \cdots + n_K; \theta_j > 0,$$

and

$$\theta = \sum_j^K \theta_j.$$

A Jump Markov Process' Stationary State then follows:

$$\pi(\mathbf{n}) w(\mathbf{n}, \mathbf{n} - \mathbf{e}_i + \mathbf{e}_j) = \pi(\mathbf{n} - \mathbf{e}_i + \mathbf{e}_j) w(\mathbf{n} - \mathbf{e}_i + \mathbf{e}_j, \mathbf{n}), \tag{12.11}$$

$$\pi(\mathbf{n}) = \frac{w(\mathbf{n} - \mathbf{e}_i + \mathbf{e}_j, \mathbf{n})}{w(\mathbf{n}, \mathbf{n} - \mathbf{e}_i + \mathbf{e}_j)} \pi(\mathbf{n} - \mathbf{e}_i + \mathbf{e}_j). \tag{12.12}$$

Hence we have the stationary distribution:

$$\pi(\mathbf{n}) = \frac{n!}{\theta^{[n]}} \prod_{i=1}^{K} \frac{\theta_i^{[n_i]}}{n_i!} \tag{12.13}$$

where

$$\theta^{[n]} = \theta(\theta+1)\cdots(\theta+n-1). \tag{12.14}$$

12.4.2 A General Urn Process

Suppose that balls (or agents) and boxes (or urns) are both indistinguishable. We then have a partition vector:

$$\mathbf{a} = (a_1, a_2, \ldots, a_n).$$

a_i is the number of boxes containing i balls. The number of balls is:

$$\sum_{i=1}^{n} i a_i = n.$$

The number of categories is:

$$\sum_{i=1}^{n} a_i = K_n.$$

K_n is the number of occupied boxes.

The number of configurations then is

$$N(\mathbf{a}) = \frac{n!}{\prod_{j=1}^{n} (J!)^{a_j} a_j!} = \frac{n!}{(1!)^{a_1}(2!)^{a_2}\cdots(n!)^{a_n} a_1! a_2! \cdots a_n!}.$$

12.4.2.1 A New Type Entry in an Urn Process

Let **a** be a state vector. Suppose that one new type agent enters an empty box. We then have the equation:

$$w(\mathbf{a}, \mathbf{a}+\mathbf{e}_1) = \frac{\theta}{n+\theta}. \tag{12.15}$$

Suppose also that an agent enters a cluster of size j. Adding one to size $j+1$ while reducing one from size j:

$$w(\mathbf{a}, \mathbf{a} + \mathbf{e}_{j+1} - \mathbf{e}_j) = \frac{ja_j}{n+\theta}. \tag{12.16}$$

On the contrary, suppose that an agent leaves a cluster of size j. Adding one to size $j-1$ while reducing one from size j:

$$w(\mathbf{a}, \mathbf{a} - \mathbf{e}_j + \mathbf{e}_{j-1}) = \frac{ja_j}{n}. \tag{12.17}$$

We then have Ewens' Sampling Formula

$$\pi(\mathbf{a}) = \frac{n!}{\theta^{[n]}} \prod_{j=1}^{n} \left(\frac{\theta}{j}\right)^{a_j} \frac{1}{a_j!} \tag{12.18}$$

where

$$\sum_{j=1}^{n} ja_j = n; \sum_{j=1}^{n} a_j = K_n.$$

12.4.2.2 The Probability that the Number of Clusters is *k*

Let the probability that the number of clusters is k be the sum of a newcomer who comes in a new cluster with probability $\frac{n}{n+\theta}$ and a newcomer who comes in an existing cluster with probability $1 - \frac{n}{n+\theta}$:

$$q_{n,k} := \Pr(K_n = k|n), \tag{12.19}$$

$$q_{n+1,k} = \frac{n}{n+\theta} q_{n,k} + \frac{\theta}{n+\theta} q_{n,k-1}. \tag{12.20}$$

In this case, there are the boundary conditions:

$$q_{n,1} = \frac{(n-1)!}{\theta^{[n]}}, \tag{12.21}$$

$$q_{n,n} = \frac{\theta^n}{\theta^{[n]}}. \tag{12.22}$$

It then follows:

$$q_{n,k} = \frac{\theta^k}{\theta^{[n]}} c(n,k) \tag{12.23}$$

where

$$\begin{aligned}
c(n+1,k) &= nc(n,k) + c(n,k-1), \\
\theta^{[n+1]} &= \sum_{m=0}^{n} s(n,m)\theta^m = \theta(\theta+1)(\theta+2)\cdots(\theta+n-1)(\theta+n) \\
&= \theta^{[n]}(\theta+n) = \theta \cdot \theta^{[n]} + n \cdot \theta^{[n]}, \\
\theta \cdot \theta^{[n]} &= \sum_{m=0}^{n} s(n,m)\theta \cdot \theta^m = \sum_{m=0}^{n} s(n,m)\theta^{m+1} \\
&= \sum_{m-1=0}^{n} s(n,m-1)\theta^m = c(n,k-1), \\
n\theta^{[n]} &= n\sum_{m=0}^{n} s(n,m)\theta^m = nc(n,k).
\end{aligned}$$

The final equation is called sign-less Stirling Number of the first kind.

12.4.3 Pitman's Chinese Restaurant Process

Suppose that there are an infinite number of round tables in the Chinese restaurant that are labeled by an integer from 1 to n. The first customer, numbered 1, takes a seat at the table numbered 1. Suppose that the customers from No.1 to No. k in turn take their seats at their tables from No.1 to No. k. Here the c_j customers take their seats at the j-th table (Pitman 1995; Yamato and Sibuya 2000, 2003).

Now the new arrival comes! The next arriving customer has two options: either to take a seat at the k-th table by the probability

$$\frac{\theta + k\alpha}{\theta + n}$$

or to take a seat at the table j, one of the remaining tables $j = 1, \ldots, k$ by the probability

$$\frac{c_j - \alpha}{\theta + n}.$$

Here two parameters, θ and α are used. Thus the solution:

$$\frac{n!\theta^{[k:\alpha]}}{\theta[n]} \prod_{j=1}^{n} \left(\frac{(1-\alpha)^{[j-1]}}{j!} \right)^{c_j} \frac{1}{c_j!} \tag{12.24}$$

where

$$\theta^{[j]} = (\theta + 1) \cdots (\theta + j - 1),$$
$$\theta^{[j:\alpha]} = (\theta + \alpha) \cdots (\theta + (j - 1)\alpha).$$

Ewens' sampling formula (Ewens 1972) gives the invariance of the random partition vectors under the properties of exchangeability and size-biased permutation. The Ewens sampling formula is the case with one parameter; in a special case of two-parameter Poisson–Dirichlet distributions:

$$\frac{n!}{\theta^{[n]}} \prod_{j=1}^{n} \left(\frac{\theta}{j}\right)^{a_j} \frac{1}{a_j!}.$$

In the case of the two-parameter Poisson–Dirichlet model, there will be a non-averaging system in the limit.

12.5 An Application of the Two-Parameter Poisson–Dirichlet Model to Avatamsaka

12.5.1 Non-Self-Averaging

Aoki and Yoshikawa (2006) have shown that the two-parameter Poisson–Dirichlet models are qualitatively different from the one-parameter version. An additional parameter could then generate non-self-averaging, even if there were a situation of self-averaging in the one-parameter model (Aoki and Yoshikawa 2007, 6). An urn state is self-averaging if the number of balls in each urn could eventually be convergent, on average, in the following sense (Aoki and Yoshikawa 2007, 4): "Non-self averaging" means that a size-dependent (i.e., "extensive" in physics) random variable X of the model has the coefficient of variation that does not converge to zero as model size goes to infinity. The coefficient of variation *C.V.* of an extensive random variable, *X*, defined by

$$\text{C.V.}(X) = \frac{\sqrt{\text{variance}(X)}}{\text{mean}(X)}$$

is normally expected to converge to zero as model size (e.g. the number of economic agents) goes to infinity. In this case, the model is said to be "self-averaging".

12.5.2 Aoki's Application

An application of the two-parameter Poisson–Dirichlet model to an economic system (Aoki and Yoshikawa 2007) shows a growing economy of multiple sectors where the waves of innovations arrive stochastically. Innovation, when it occurs, either raises productivity in one of the existing sectors, or creates a new sector. Thus, the number of sectors is not given, but increases over time.

By the time the n-th innovation occurs, the total of K_n sectors are formed in the economy where the i-th sector has experienced n_i innovations ($i = 1, 2, \ldots, K_n$). By definition, the following equality holds:

$$n_1 + n_2 + \cdots + n_k = n$$

when $K_n = k$. If the n-th innovation creates a new sector (sector k), then $n_k = 1$.

12.5.2.1 Between Different Spillovers

The Avatamsaka game is characterized by the positive spillovers for both players; larger or smaller spillovers all preserve Avatamsaka characteristics, of course. A typical change of spillover size is illustrated in Fig. 12.9. As stated in Sect. 12.1.4, our Avatamsaka game could always be defined as the contours, if the various different payoffs are allowed to be introduced. Different contours correspond to different payoffs of the Avatamsaka game. Given the different payoff states, players can wander among different states. So we can describe a kind of dynamic equilibrium.

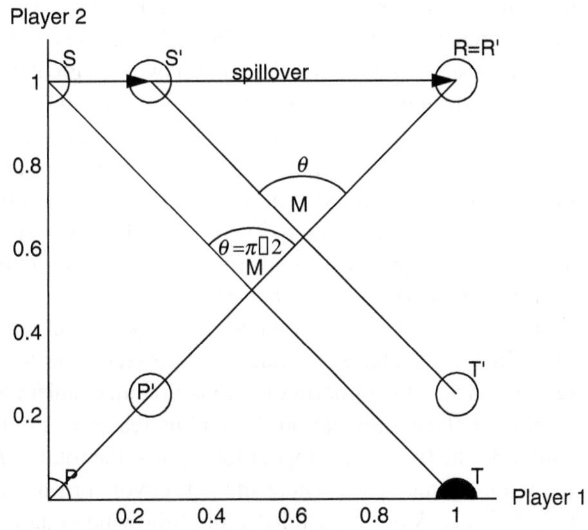

Fig. 12.9 Spillover

Fig. 12.10 Various spillovers

We can measure a difference of any two spillovers by

$$\frac{\eta_1}{\eta_2} = \frac{SR}{S'R'}.$$

Incidentally, we note that we can get a greater spillover structure in a Prisoner's Dilemma game by simply extending the *PR* line (Fig. 12.10).

Given a particular spillover, the expected total gain in the urn will increase, if the level of cooperation in the game increases. It is natural that more cooperative players could bring about a higher average size gain in the given box. This coincides with the previous assumption (*Assumption 2*).

12.5.3 An Application to Avatamsaka Game

Suppose there are many different urns with various different payoff matrices, each of which has its own size spillover. A different spillover in the Avatamsaka game may change the inclinations of players' reactions. However, these inclinations are not necessarily symmetrical. An urn with a greater spillover might sometimes be more attractive for a cooperative player in the context of a high level of cooperation, because the player could earn greater gains. A cooperative player in the context of a low level of cooperation might enter an urn with a smaller spillover than he would with a much higher level of cooperation. He can change his mind to become a defecting player. There may thus be various players' plans for their strategies. Any player depends on his state, which may change, while an urn, that is, a payoff matrix occurs stochastically.

Thus, any urn must be random for players, and any player must be random for urns. In this Avatamsaka game, size effects due to different spillovers may be ignored. A specific or particular transition mechanism need not be assumed ex ante.

Now replace "innovations" with "increases of the level of cooperation" in this game. By the time n-th cooperation occurs, the total of K_n payoff urns are formed in the whole game space where the i-th payoff urn has experienced n_i cooperations ($i = 1, 2, \ldots, K_n$). By definition, the following equality holds:

$$n_1 + n_2 + \cdots + n_k = n \tag{12.25}$$

when $K_n = k$. If n-th cooperation creates a new payoff matrix (urn k), then $n_k = 1$. It is noted that

$$n = \sum_j j a_j(n). \tag{12.26}$$

So there are a finite number of urns into which various types of payoff matrices are embedded: 1, ..., K.

In this environment, we can have n inventions to increase the level of cooperation x_i. In other words, the level of cooperation in urn i may grow due to stochastic multiple inventions occurring in this urn.

Assumption 3 cooperation accelerates cooperation, i.e., the larger the number of cooperators, the larger the number of cooperators and/or the greater the total gain in the urn will be.

Assumption 4 due to an Avatamsaka property, a player can compare the situations between the urns given to him by normalizing his own gain.

We then assume a particular type growth:

$$x_i = \eta_i \gamma^{n_i}, \eta_i > 0, \gamma > 1, \text{for } i = 1, ..., k. \tag{12.27}$$

First of all, the part γ^{n_i} reflects *Assumption 3*.

As shown above, η_i indicates an element of the set of different spillovers:

$$E = (\eta_1, \cdots, \eta_n).$$

$a_j(n)$ is the number of urns where j inventions have occurred. Here $a_j(n)$ is an element of the partition vector $a(n)$. K_n can then be expressed as

$$K_n = \sum_j^n a_j(n). \tag{12.28}$$

Expanding the exponential $\exp(n_i \ln \gamma)$ and rounding the remaining terms except for the first two, we obtain the next approximation:

$$\gamma^{n_i} \approx 1 + \ln(\gamma) n_i.$$

Hence it follows:

$$x_i = \eta_i + \eta_i \ln(\gamma) n_i. \tag{12.29}$$

Fig. 12.11 A stochastic interaction

Due to Assumption 4, we normalize x_i by the use of η_i (Fig. 12.11).

$$\tilde{x}_i = \frac{x_i}{\eta_i}.$$

Here we then define:

$$X_n = \sum_{i}^{K_n} \tilde{x}_i. \tag{12.30}$$

This shows the aggregate behavior of cooperation dynamics, i.e., cluster urn dynamics. Thus, from these (12.27)–(12.30) in the above, we obtain

$$X_n \approx K_n + \beta \sum_{j}^{n} j a_j(n), \tag{12.31}$$

where $\beta = \ln(\gamma) > 0$. It then turns out that X_n depends on how cooperation occurs.

12.5.4 A Result

We have just transformed our Avatamsaka game form into the same equation (12.31) in essentially the same context as Aoki and Yoshikawa (2007). Thus, by the same reasoning, we could conclude the same result. Hence we have the following proposition[11]:

[11]Aoki and Yoshikawa (2007, p. 6) argued the economic meaning of non-self-averaging as follows: the notion of non-self-averaging is important because non-self-averaging models are sample dependent, and some degree of impreciseness or dispersion remains about the time trajectories even when the number of economic agents goes to infinity. This implies that a focus on the mean path behavior of macroeconomic variables is not justified. It, in turn, means that sophisticated optimization exercises that provide us with information on the means have little value.

Proposition 3. *In the two-parameter Poisson–Dirichlet model, the aggregate cooperation behavior X_n is non-self-averaging.* (For the proof, see Aoki and Yoshikawa 2007, pp. 6–10.) □

12.6 Concluding Remarks

There is a particular type of self-fulfilling prophecy that is like the Oedipus effect. According to Robert K. Merton, this effect is that a true prophetic statement – a prophecy declared as truth when it is not – may sufficiently influence people, either through fear or logical confusion, so that their reactions ultimately fulfill the false prophecy. This implies that the collective macro-state of various social orders (order parameter) can be averaged over its parts.

We apply this self-fulfilling rhetoric to our rational expectation hypothesis. This hypothesis may be self-fulfilling either through fear or logical confusion, of course, but this prophecy must be faced with some logical failure. It could be verified in the framework of complex dynamics that this hypothesis only referred to a unilateral direction of the whole dynamics, in which a single rational individual had to contribute to the rational macro-state of the economy. Actually, we need the other aspect of the full feedback: "Its order parameters strongly influence the individuals of the society by orienting (enslaving) their activities and by activating or deactivating their attitudes and capabilities". (Mainzer 2007a, p. 395). This is just the slaving principle as a whole elucidated as synergetics by Hermann Haken (Weidlich 2002, 2006, 2007). This dynamic thus is encompassed by critical values, outside of which the system falls into an unstable situation. So the prophecy to be fulfilled might be circumvented. In our new setting of the Avatamsaka game, it has turned out that we often are faced with non-convergent, unstable situations in the face of some particular types of interactive chances.

References

Akiyama E, Aruka Y (2006) Evolution of reciprocal cooperation in the Avatamsaka game. In: Namatame A, Kaizoji T, Aruka Y (eds) The complex networks of economic interaction. Springer, Heidelberg, pp 307–321

Aoki M (1996) New approaches to macroeconomic modeling: evolutionary stochastic dynamics, multiple equilibria, and externalities as field effects. Cambridge University Press, New York

Aoki M (2002) Modeling aggregate behavior and fluctuations in economics: stochastic views of interacting agents. Cambridge University Press, New York

Aoki M (2008) Dispersion of growth paths of macroeconomic models in thermodynamic limits: two parameter Poisson-Dirichlet models. J Econ Interact Coord 3(1):3–14

Aoki M, Yoshikawa H (2006) Reconstructing macroeconomics: a perspective from statistical physics and combinatorial stochastic processes. Cambridge University Press, Cambridge, NY

Aoki M, Yoshikawa H (2007) Non-self-averaging in macroeconomic models: a criticism of modern micro-founded macroeconomics. Economics Discussion Papers (http://www.economics-ejournal.org) 2007-49 November 26

Arthur BW (1994) Increasing returns and path dependence in the economy. University of Michigan Press, Ann Arbor

Arthur BW, Ermoliev YM, Kaniovski YM (1987) Path-dependent processes and the emergence of macro-structure. Eur J Oper Res 30:294–303

Aruka Y (2001) Avatamsaka game structure and experiment on the Web. In: Aruka Y (ed) Evolutionary controversies in economics. Springer, Tokyo, pp 115–132

Aruka Y (2001) Avatamsaka game experiment as a nonlinear Polya Urn Process. In: Terano T, Namatame A et al (eds) New frontiers on artificial intelligence. Springer, Berlin, pp 153–161

Aruka Y (2004) How to measure social interactions via group selection? a comment: cultural group selection, coevolutionary processes, large-scale cooperation. J Econ Behav Org 53 (1):41–47

Aruka Y (2007) The moral science of heterogeneous economic interaction in the face of complexity. In: Theodor Leiber (Hg) Dynamisches Denken und Handeln Philosophie und Wissenschaft in einer komplexen Welt, Festschrift fuer Klaus Mainzer zum 60. Geburtstag S.Hirzel Verlag Stuttgart, pp 171–183

Aruka Y (2009) Book Review: Klaus Mainzer, Der kreative Zufall: wie das Neue in die Welt kommt (The creative chance. how novelty comes into the world (in German)). C.H. Beck, München, 2007, 283p. Evol Inst Econ Rev 5(2):307–316

Bowles S (2004) Microeconomics: behavior, institutions, and evolution. Princeton University Press, Princeton

Canfield J, Hansen MV (2001) Chicken soup for the soul: 101 stories to open the heart and Rekindle the spirit (Chicken Soup for the Soul). Health Communications, Arlington, VA

Ewens WJ (1972) The sampling theory of selectively neutral alleles. Theor Popul Biol 3:87–112

Flajolet P, Gabarro J, Pekari H (2005) Analytic urns annals of probability. Ann Probab 33 (3):1200–1233

Fujiwara Y (2008) Book Review: Masanao Aoki and Hiroshi Yoshikawa, Reconstructing macroeconomics – a perspective from statistical physics and combinatorial stochastic processes. Cambridge University Press, 2007, 352 p. Evol Inst Econ Rev 4(2):313–317

Henrich J (2004) Cultural group selection, coevolutionary processes and large-scale cooperation. J Econ Behav Org 53:3–36

Hildenbrand W (1994) Market demand. Princeton University Press, Princeton

Mainzer K (2007a) Thinking in complexity. The computational dynamics of matter, mind, and mankind, 5th edn. Springer, New York

Mainzer K (2007b) Der kreative Zufall: wie das Neue in die Welt kommt (The Creative Chance. How Novelty Comes into the World). C. H. Beck, München

Mandelbrot BB, Hudson RL (2004) The (Mis)behavior of markets: a fractal view of risk, ruin and reward. Basic Books, New York

Mitropoulos A (2004) Learning under minimal information: an experiment on mutual fate control. J Econ Psychol 22:523–557

Nowak MA, Sigmund K (1993) A strategy of win-stay lose-shift that outperforms tit-for-tat in the Prisoner's Dilemma game. Nature 364:56–58

Pitman J (1995) Exchangeable and partially exchangeable random partitions. Probab Theory Relat Fields 12:145–158

Pitman J (2002) Lecture Notes of the Summer School on Probability, St. Flour, France (forthcoming from Springer)

Price GR (1970) Selection and covariance. Nature 227:520–521

Price GR (1972) Extension of covariance selection mathematics. Ann Hum Genet 35:485–490

Tanimoto J (2007a) Promotion of cooperation by payoff noise in a 2 times 2 game. Phys Rev E 76:0411301–0411308

Tanimoto J (2007b) Does a tag system effectively support emerging cooperation. J Theor Biol 247:756–764
Tanimoto J, Sagara H (2007) Relationship between dilemma occurrence and the existence of a weakly dominant strategy in a two-player symmetric game. Bio Syst 90:105–114
Thibaut JW, Kelley HH (1959) The social psychology of groups. Wiley, New York
Weidlich W (2002) Sociodynamics: a systematic approach to mathematical modeling in the social sciences. Taylor and Francis, London
Weidlich W (2006) Intentions and principles of sociodynamics. Evol Inst Econ Rev 2(2):161–166
Weidlich W (2007) Laudatio inofficialis für Prof. Dr. Dr. H. C. Mult, Hermann Haken anlasslich seines 80. Geburtstages (mimeo)
Yamato H, Sibuya M (2000) Moments of some statistics of Pitman sampling formula. Bull Inform Cybernet 6:463–488
Yamato H, Sibuya M (2003) Some topics on Pitman's probabilistic partition. Stat Math 51:351–372

Appendix
Klaus Mainzer Der kreative Zufall: wie das Neue in die Welt kommt (The Creative Chance. How Novelty comes into the World, German),[1] C.H. Beck, München, 2007, 283pp.

Yuji Aruka

A.1 Contributions and His Short Curriculum Vitae of Prof. Dr. Klaus Mainzer

A.1.1 *His Contributions from Thinking in Complexity*

Prof. Dr. Mainzer is now quite familiar to many as the author of *Thinking in Complexity* which has been a quite successful and an influential book all over the world is now in the particularly in the field of complex system and philosophy of science. This book is now in the enlarged 5th edition; the first edition appeared in 1994.

Prof. Mainzer in this enlarged version (Mainzer 2007a) insightfully summarized the history of philosophy and ethics. He recognized that Kant's ethics and Anglo-American utilitarianism are normative demands to judge our actions. In other words, the utilitarian principle of happiness can be regarded as Kant's material completion. According to Mainzer, however, Hegel argued more smartly, i.e., "he distinguished between the subjective morality and subjective reason of individuals and the objective morality and subjective reason of institutions in a society", (Mainzer 2007a, p. 405), though he did not have any idea of nonlinear integration due to emerging chances and their interaction. Of course, it is noted that his historical belief was maliciously misused by totalitarian politician later. It must however be true that the same result applied to neoclassical economics. Neoclassical economics certainly fantasized the so-called linear assumption that "coupling the dynamics of free markets and democracy would automatically lead to a community of modernized, peace-loving nations with civic-minded citizens and consumers". (Mainzer 2007a, p. 408) This was

[1]Here "chance" corresponds to "Zufall" in German but the two are not precisely the same thing. Mainzer himself translated Zufall into "chance" when he wrote the English title.

Y. Aruka
Faculty of Commerce, Chuo University, Higashinakano, Hachioji, Tokyo 192-0393, Japan
e-mail: aruka@tamacc.chuo-u.ac.jp

DOI 10.1007/978-3-7908-2618-0,

a terrible error after the fall of Berlin wall. We should now obtain new insight from nonlinear sciences and complexity thinking, so that we are ready to argue positively Hegel's methodology in the new methodological view. In our secular world, "Good intentions may lead to bad effects without their subjective intentions. Hegel called it a stratagem of reason (List der Vernunft)". (Mainzer 1994–1995, p. 408) Traditional economics has excessively overestimated subjective intentions without any reference to the objective morality and subjective reason of institutions in a society. It then continues in economics to suffer from standing on the poor philosophy.

Thus Mainzer has begun to undertake a new project. At the very end of the volume of the enlarged version (Mainzer 1994 [2007, pp. 411–412]), Mainzer has completed his book by stating the next phrases:

> Immanuel Kant summarized the problem of philosophy in the three famous questions: what can I know?; what must I do?; what may I hope? The first question concerns epistemology with the possibilities and limitations of our recognition. ... The second question concerns ethics and the evaluation of our actions. ... Kant's last question "What may we hope" concerns the Greatest Goods, which has traditionally been discussed as summum bonum in the philosophy of religion. At first glance, it seems to be beyond the theory of complex system. ... But when we consider the long sociocultural evolution of mankind, the greatest good that people have struggled for has been the dignity of their personal life. This does not depend on individual abilities, the degree of intelligence, or social advantages acquired by the contingencies of birth. It has been a free act of human self-determination in a stream of nonlinearity and randomness in history. We have to project *the Greatest Good* (italics by the reviewer) on an ongoing evolution of increasing complexity.

From this citation, we immediately realize that the author's philosophical attempt must be a very attractive integration of Anglo-American minded empiricism with German idealism. This attitude is just one what the reviewer expected for a long time since starting the study of economics. Thus we must reconsider "a free act of human self-determination in a stream of nonlinearity and randomness in history". It is easy to realize that individuals never rely upon their own a priori preferences only. On the other hand, we cannot neglect "chances" independently of "individual abilities" (including learning abilities) and "circumstances" which must be forced to accept. Things like novelties emerging either from nature or society could never happen without intermediation of chances. Thus we need to investigate how chances may intermediate emergences of novelties. This is the reason why Mainzer wrote the next book titled: The *Creative Chance. How Novelty comes into the World* (Mainzer 2007b).

A.1.2 *Curriculum Vitae of Prof. Dr. Mainzer*

Before reviewing *The Creative Chance*, we would like to add to more on the author's personal profile. For convenience, only works of single authorship are given at the end of this review.

Prof. Dr. Klaus Mainzer, who was born in 1947. At university, he studied mathematics, physics, and philosophy. After completion of Ph.D. in 1973 and the Habilitation in philosophy in 1979, he served as assistant and lecturer (1973–1979) at the University of Munster he was awarded a Heisenberg scholarship in 1980. His professional appointments were as follows: 1981–1988, professor for philosophy (foundations and history of science) at the University of Constance; vice-president of the university of Constance (1985–1988); 1988–2008, chair for philosophy of science, dean (1999–2000), director of the Institute of Philosophy (1989–2008) and of the Institute of Interdisciplinary Informatics (1998–2008) at the University of Augsburg. He now is currently professor of Technische Universität München (Technical University, Munich). He is the chair for philosophy of science and technology, the director of the Carl von Linde-Academy and a member of the advisory board of the Institute of Advanced Study at the Technical University of Munich since 2008. He is also a member of several academies and interdisciplinary organizations (e.g., European Academy of Sciences/Academia Europaea in London, Daimler-Benz Foundation), and president of the German Society of Complex Systems and Nonlinear Dynamics (since 1996).[2] Additionally, he currently is the chairman of the scientific board of the Japanese–German Society of Integrative Science (Leibniz-Association e.V.). This society whose activity is based on the Leibniz-Association has been officially approved as an academic society by German Government in 2008 after the 5 years preparatory activities.[3]

The Technische Universität München has produced 14 Nobel prize laureates since 1927 to 2008. Professor Mainzer honourably has served as the chair for philosophy of science and technology since April 2008. We feel this is a matter of course based on his many splendid achievements of his academic career. He is really one of the most influential leading scholars on the philosophy of science and complexity thinking in the world.

A.2 How Novelty Comes into the World

A.2.1 *The Philosophy of Chances and Contingencies*

In the book of Mainzer (2007b) the term of Zufall (Chance) appears in the book title. According to Mainzer (2007b, p. 32), Kant definitely identified "Contingency" with "Zufall" in German. He furthermore distinguished contingencies between

[2]He is also the co-editor of international scientific journals: e.g., International Journal of Bifurcation and Chaos in Applied Sciences and Engineering (Singapore: World Scientific); Advanced Science Letters (American Scientific Publishers); Kant-Studies (Berlin: De Gruyter).

[3]Prof. Dr. Roman Herzog, the former president of Federal Republic of Germany, was inaugurated as the president of this society. For the readers who are interested in this society, see the URL: http://www.integrative-wissenschaft.de/index.html

more empirical, more logical, and more intelligible one. An empirical contingency which depends on a certain cause is meant as the "Event". Sun irradiation is not a necessary condition for heating a stone. While a stone necessarily falls on the basis of the gravity law to the earth at any moment. Seemingly we have opposite observations. Concept may fail to catch all the events around it. A property (attribute), that does not follow from the definition of a concept (category), is called logically "Chance".

Mainzer continues his argument further (Mainzer 2007b, p. 32). The Scottish philosopher David Hume (1711–1776) doubts not only knowledge independently from the human perception but also the necessary categories of the thinking like causal law. Causes and effects rather are perceptions which are mutually associated on the basis of their repeated occurrences in the human consciousness. Thunder comes after lightning repeatedly, so these perceptions are associated in the memory and lead to an expectation as well as a belief of the consciousness.

Hume's epistemological skepticism is radical. He doubts not only deterministic causal laws of nature and thinking. He also doubts the blind coincidence in nature since we can never know with certainty whether the observation of an event, whose cause we do not know, does not have a cause. However, our ignorance of a cause itself is a fact which influences our beliefs, our assumptions and expectations of the events. These beliefs becomes measurable in the probabilities which we can assign to the events on the basis of the frequency. Therefore its legitimacy is not necessarily deterministic but has the purpose to predict the events *with certainty*. We must rather focus on the degrees of the regularity, with which the events can be arriving. In the mechanics, as Hume explains in details, no exceptions are known. In the rules of medicine for example, however, the degree of the regularity decreases while the number of the exceptions increases. Prognoses become possible with different degrees of the certainty.

Thus Mainzer concluded that David Hume was the pioneer who became a consistent epistemological theorist and introduced modern thinking in probability.

A.2.2 *The Transition from* Tyche *and* Fortuna *to* Kairos

In Chapter 1 Mainzer (2007b), Mainzer noticed the historical transition from *Tyche* (τύχη)[4] in ancient Greek who controlled the fate of humans as to whether their luck was good or bad to *Fortuna* in ancient Rome who was the goddess of good fortune. In this transition, the bad luck of *Tyche* was dropped. *Fortuna* however does not always guarantee fortunes for individuals. *Oh Fortuna, velut luna statu variabilis*

[4]In Greek mythology, *Tyche* (Τύχη) was the goddess of fortune. She was chosen by several ancient Greek cities as the protector that governed their fortune and prosperity.

("Ah *Fortuna*, always changeable like the moon").[5] That is to say, she appears veiled and blind, so she is regarded as *capriciousness*. The darkness in the lives of individuals could not be removed by these goddess. The Christian tradition rejected the idea of goddesses of fortune and fate, and *Kairos*, instead of *Tyche* and *Fortuna*, became highlighted in the new context. Individuals can prepare for the moment which might bring them opportunities. *Kairos* who was the youngest son of *Zeus* was the personification of the right moment, the opportune moment. We introduce two parables from Genesis. Noah from his divine intuition could predict the great flood, so he then was able to prepare his ark, to protect himself. Joseph in Egypt realized from Pharaoh's dream that the prosperity could not last. He warned that the state should reserve food stocks before the famine came. The state must behave aperiodic, indeed. Thus we must actually not pass over the moment of favorable decision. According to Mainzer (2007b), this methodological view provided the turning point from ancient philosophies where the personifications of decision moment were eliminated.

Random single events cannot be predicted respectively, but the signs of the moment might be recognized in which they could take place. This is what we have learned from the idea of *Kairos*. By relying on Mandelbrot's metaphor (Mandelbrot and Hudson 2004), we can now elucidate both of the Noah and the Joseph effect in the device of fractal geometry, as Mainzer chose to describe Chap. 7 titled "Chance in Culture, Economy and Society".

A.2.3 *Mandelbrot's Parable: The Joseph Effect and the Noah Effect*

We now return Mainzer's earlier work *Thinking in Complexity* to illustrate the key issues. As we already discussed above, "we need new models of collective behavior depending on the different degrees of our individual faculties and insights. *In short* [italics by the reviewer]: The complex system approach demands new consequences in epistemology and ethics. Finally, it offers a chance to handle chaos and randomness in a nonlinear world and utilize the creative possibilities of synergetic effects". (Mainzer 2007a, p. 15) "One of the deepest insights into a complex system is the fact that even complete knowledge of microscopic interactions does not guarantee prediction of the future". (Mainzer 2007a, p. 395).

On the contrary, in classical physics, changes on a course run continuously. Similarly in 1890, Alfred Marshall, the leading neo-classical economist, inherited from Gottfried Leibniz the motto of *Natura non facit saltus* ("nature does not make (sudden) jumps") which has been a principle of natural philosophies since Aristotle's time (Mainzer 2007b, p. 219). The market thus never made "any jump".

[5]In Roman mythology, *Fortuna* was the goddess of fortune and equivalent to the Greek goddess *Tysche*.

In the light of nonlinear and random dynamics, however it is easily verified that this statement is false, as proven in detail by Mainzer (2007a, b).

We have a particular type of self-fulfilling prophecy like *the Oedipus effect*. According to this effect interpreted by Robert K. Merton, a true prophetic statement – a prophecy declared as truth when it is not – may sufficiently influence people, either through fear or logical confusion, so that their reactions ultimately fulfill the false prophecy. This implies that the collective macrostate of social various orders (order parameter) can be averaged over its parts. We apply this self-fulfilling rhetoric to our rational expectation hypothesis. This hypothesis might be self-fulfilling either through fear or logical confusion, of course. But this prophecy must be faced with some logical failure. It would be verified in the framework of complex dynamics that this hypothesis only referred to a unilateral direction of the whole dynamics in which direction a single rational individual had to contribute to the rational macrostate of economy. Actually we moreover need the other aspect of the full feedback: "Its order parameters strongly influence the individuals of the society by orientating (enslaving) their activities and by activating or deactivating their attitudes and capabilities". (Mainzer 2007a, p. 395) This is just *the slaving principle* as a whole elucidated as *synergetics* by Hermann Haken (1977) and Wolfgang Weidlich (2002, 2007).[6] This dynamics thus is encompassed by critical values, outside of which the system falls into an instable situation. So the prophecy to be fulfilled might be circumvented.

In parallel with the final chapter on ethics and society,[7] in *Thinking in complexity*, *The Creative Chance* also has the similar chapter titled: "Chance in Culture, Economy and Society". Here Mainzer has referred to the two effects of economic time series. These effects were called the Joseph Effect and the Noah Effect (Mandelbrot and Hudson, 2004) by Mandelbrot, a great fractal geometrician who was inspired by the research of Hurst (1951)[8] on the fluctuations of the level of the Nile River. The Noah Effect states that change happens in discrete jumps. The Joseph effect states that some things tend to persist. These two effects push the world in different directions (Gleick 1987, pp. 92–94). The parable which Mandelbrot employed may be read as follows:

- The Joseph effect describes persistence. Joseph's interpretation of Pharaoh's dream of seven fat cows and seven gaunt ones to mean that there would be 7 prosperous years followed by seven lean ones. He discovered that *trends tend to persist*; that is, if a place has been suffering drought, it's likely it will suffer more of the same. In other words, *things tend to stay the way they've been recently*.

---

[6]This principle was first discovered by the research on laser light.

[7]The final chapter of Mainzer (2007a) is Chapter 8 titled: "Epilogue on Future, Science and Ethics".

[8]Harold Edwin Hurst (1880–1978) was a British hydrologist. Hurst's (1951) study on measuring the long-term storage capacity of reservoirs documented the presence of long-range dependence in hydrology.

- The Noah Effect describes discontinuity. The source is the story of Noah's ark in the Deluge (the Great Flood). Given a situation, something significant will trigger its collapse. As Mandelbrot then envisaged it, when something changes, it can change abruptly. The stock market is a very nice example for this.[9]

At the time when something significant triggers its collapse, *Kairos* may appear because Kairos is just the right moment to provide for every eventuality of both types: Noah and Joseph. We can learn here that science never be separated from ethics.

It is quite interesting to know that a geometric science of fractal, one of the new important methods generating complex dynamics provides a new ethics. So we reproduce Figure 38 of Mainzer (2007b, p. 203) which is cited from Mandelbrot and Hudson(2004). These figures respectively are depicted by the different Hurst exponential coefficients H. We obtain the upper panel (Fig. A.1.a) when H is smaller than 0.5 ($H < 0.5$). The middle (Fig. A.1.b) appears when $H = 0.5$. The bottom (Fig. A.1.c) when H is greater than 0.5 ($H > 0.5$). The upper represents an instable movement due to the Noah effect. The middle represents Brownian movement. The bottom represents a stable movement due to the Joseph effect.

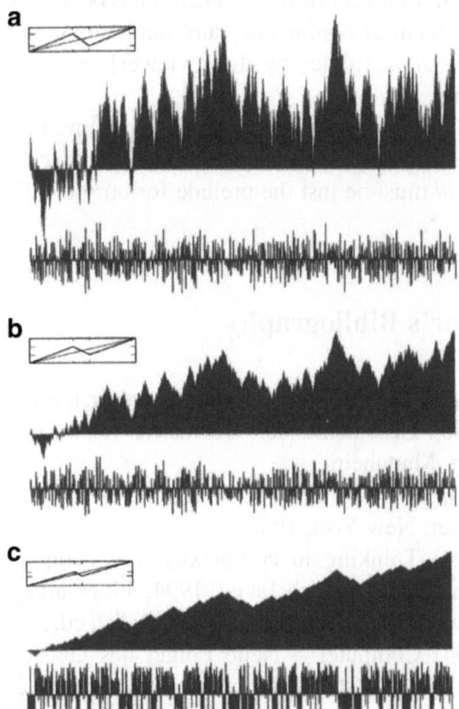

Fig. A.1 Fractal generating random curves

[9]According to James Gleick (1987), the Noah and Joseph Effects push in different directions, but they add up to this: trends in nature are real, but they can vanish as quickly as they come. This was cited from Roger von Oech's blog: http://blog.creativethink.com/2007/07/the-joseph-effe.html

A.2.4 *The New Project for Moral Science in the New Era*

In summing up, we finally paraphrase the end of the book (Mainzer 2007b, p. 230): Without chance nothing new happens. The events and outcomes do not always favour us. These may cover from viruses and deceases to crazy markets and men involved into felonious activities. Chance cannot be indeed calculated and controlled. We could however analyse and understand the laws of its system. Given the suitable side conditions, the assumptions can then be created under which random events and synergetic effects dissolved by themselves. Systems can then be brought by self-organizing in our sense and by creating the results which favour us, whether they are "quanta, organisms, computer" in nature, "technique and medicine", or "mens's intelligence on the markets" in culture and society. Favourable results are not always guaranteed us. No guarantee represents the price of chance.

However chance then makes sense for us. The creative chance will be brought about from the blind. *Kairos* then appears while *Tyche* disappeared.

We cite again from Mainzer (2007a, p. 412). It has been a free act of human self-determination in a stream of nonlinearity and randomness in history. We have to project *the Greatest Good* [italics by the reviewer] on an ongoing evolution of increasing complexity.

We shall actually need such *Kairos* in order to procure the greatest good in the increasingly perplexing world of the new era. *The Creative Chance. How Novelty Comes into the World* must be just the prelude for our new philosophy.

A.3 The Author's Bibliography

We for brevity mention the references of single authorship only:

Mainzer K (1980) Geschichte der Geometrie (History of Geometry), B. I. Wissenschaftsverlag, Mannheim

Mainzer K (1988, 1996) Symmetries of nature, De Gruyter 1996; *Symmetrien der Natur*, De Gruyter, New York 1988

Mainzer K (1994) Thinking in complexity. the computational dynamics of matter, mind, and mankind (English 1st ed. 1994, 5th enlarged ed. 2007) Springer, New York (Japanese ed. 1997, Chinese ed. 1999, Polish ed. 2007, Russian ed. 2008)

Mainzer K (1994) Computer – Neue Flügel des Geistes? (Computer –New Wings of the Mind?) 2. Aufl. De Gruyter, Berlin 1995 (German: De Gruyter 1st ed. 1994, 2nd ed. 1995)

Mainzer K (1995, 2002) The little book of time (English ed. Copernicus Books 2002, German: C.H. Beck 1995, 5th edition 2005, Chinese 2003, Korean 2005)

Mainzer K (1996) *Materie. Von der Urmaterie zum Leben* (Matter. From Its Origin to Life). C. H. Beck, München German: C.H. Beck 1996 (Chinese ed. 2000)

Mainzer K (1997) *Gehirn, Computer, Komplexität* (Brain, Computer, and Complexity), Springer, Berlin

Mainzer K (1999) Computernetze und virtuelle Realität. Leben in der Wissensgesellschaft (Computational Networks and Virtual Reality. Living in Knowledge Societies), Springer, Berlin

Mainzer K (2000) Hawking (Hawking. Master Thinker of Cosmology). Herder, Freiburg

Mainzer K (2003) KI – Künstliche Intelligenz. Grundlagen intelligenter Systeme (AI –Artificial Intelligence. Foundations of Intelligent Systems). Wissenschaftliche Buchgesellschaft, Darmstadt

Mainzer K (2003) Computerphilosophie (Computational Philosophy). Junius-Verlag, Hamburg

Mainzer K (2005) Symmetry and complexity. The spirit and beauty of nonlinear science. World Scientific, Singapore

Mainzer K (2007) Der kreative Zufall: wie das Neue in die Welt kommt (The Creative Chance. How Novelty comes into the World). C.H. Beck, München

Mainzer K (2008) Komplexität, UTB Profile (Uni-Taschenbücher S), Stuttgart

As for Co-author/(Co-) Editor of books, we listed in the footnote.[10]

References

Gleick J (1987) Chaos: making a new science. Viking Penguin, New York

Haken H (1977) Synergetics – an introduction: nonequilibrium phase transitions and mankind, Springer, Berlin

Hurst H (1951) Long term storage capacity of reservoirs. Trans Am Soc Civil Eng 116:770–799

[10]The following are the co-authored or (co-)edited books: H.-D. Ebbinghaus, H. Hermes, F. Hirzebruch, K. Mainzer et. al., *Numbers*, Springer, New York, 1990; German 1st ed. 1983, 3rd ed. 1992, Japanese ed. 1992, French ed. 1999; J. Audretsch and K. Mainzer (eds), Philosophie und Physik der Raum-Zeit, B. I. Wissenschaftverlag, Mannheim, 1st ed. 1988, 2nd ed. 1994; J. Audretsch and K. Mainzer (eds), Vom Anfang der Welt—Wissenschaft, Philosophie, Religion, Mythos, C. H. Beck, Munich, 1st ed. 1989, 2nd ed. 1990; J. Audretsch and K. Mainzer (eds), Wieviele Leben had Schrodingers Katze—Zur Physik und Philosophie der Quantenmechanik, B. I. Wissenschaftverlag, Mannheim, 1st ed. 1990, Spektrum, Heidelberg, 2nd ed. 1996; K. Mainzer (ed), Natur- und Geistwissenschaften—Perspektiven und Erfahrungen mit fachübergreifenden Ausbildungshalten, Springer, Berlin, 1990; E.P. Fischer and K. Mainzer, Die Frage nach dem Leben, Piper, Munich, 1990; K. Mainzer (ed), Ökonomie und Ökologie unter besonderer Berücksichtigung der Alpenregion, Haupt, Bern, 1993; K. Mainzer and W. Schirmacher (eds), Quanten, Chaos und Dämonen: Erkenntnistheoretische Aspekte der modernen Physik, B. I. Wissenschaftverlag, Mannheim,1994; K. Mainzer, A. Müller and W.G. Saltzer (eds), *From Simplicity to Complexity: Information, Interaction, and Emergence*, Vieweg, Braunschweig, 1998; K. Mainyer (ed), Komplexe Szsteme und Nichilineare Dynamik in Natur und Gessellschaft, Springer, Berlin, 1999 and so on.

Mainzer K (1994) Thinking in complexity. The computational dynamics of matter, mind, and mankind (English 1st ed. 1994, 5th enlarged ed. 2007) Springer, New York

Mainzer K (2007) Der kreative Zufall: wie das Neue in die Welt kommt (The Creative Chance. How Novelty comes into the World). C.H. Beck, München

Mandelbrot BB, Hudson RL (2004) The (Mis)behavior of markets, a fractal view of risk, ruin and reward. Basic Books, New York; Fraktale und Finanzen. Märkte zwischen Risiko, Rendite und Ruin, Piper, München (German 2005)

Weidlich W (2000) Sociodynamics: a systematic approach to mathematical modeling in the social sciences. Harwood Academic, Publishers, Amsterdam (The Gordon and Breach Publishing Group). [Reprinted by Taylor and Francis (2002); Paper edition, Dover Publications (2006); Japanese translation, Morikita Shuppan (2007)]

Weidlich W (2007) Laudatio inofficialis für Prof. Dr. Dr. h. c. mult, Hermann Haken anlässlich seines 80. Geburtstages (mimeo)

MIX
Papier aus verantwortungsvollen Quellen
Paper from responsible sources
FSC® C105338

If you have any concerns about our products,
you can contact us on
ProductSafety@springernature.com

In case Publisher is established outside the EU,
the EU authorized representative is:
Springer Nature Customer Service Center GmbH
Europaplatz 3, 69115 Heidelberg, Germany

Printed by Libri Plureos GmbH
in Hamburg, Germany